Winners & Losers
Loco Bashing Tales from the 1980s

To my parents and grandparents, who took
me on and allowed me to make so many of the
journeys within this book.

by Andy Chard

Published by Platform 5 Publishing Ltd,
52 Broadfield Road, Sheffield, S8 0XJ. England.

Printed in England by The Lavenham Press, Lavenham, Suffolk.

ISBN: 978 1 909431 75 1

Front cover: A photograph that sets the scene for this book in a number of ways. Taken at the time when its tales begin, the image also captures something of the dedication that bashers have for their beloved locomotives. On 29 September 1984 45074 passes Dawlish while in charge of the 09.10 Paignton–Newcastle. *Simon Record*

Back Cover: Class 31s could often be found on the line from Exeter to Barnstaple in the 1980s. No heat 31296 is seen at Exeter St. Davids before departing with the delayed 16.45 to Barnstaple on 6 July 1987. *Gray Callaway*

Left: During the 1980s York was a hotbed for locomotives, with regular diesel-hauled passenger and freight trains passing through and plenty of observers on the platform ends. On Sunday 26 March 1989, 47465 heads south with the 08.00 Glasgow–Penzance. *Richard Allen*

CONTENTS

ACKNOWLEDGEMENTS

The majority of the historical information within this book has been taken from my own records, which have been meticulously kept over many years. Other sources, however, have been drawn from in order to fill in some of the gaps, particularly for locomotive diagrams and workings data. Some of the sources are written, including the Loco Hauled Travel series published by Metro Enterprises, previous editions of the British Rail Passenger Timetable and various editions of Platform 5's Locomotives Pocket Book. There are an abundance of online electronic resources which have been consulted, which include _www.thebashingyears. co.uk, www.brdatabase.info, www.class37.co.uk_ and Wikipedia. Some of the most useful information has been from the many photographs and accompanying details that individuals have submitted on the great number of locomotive-based groups across social media and on the website _www.flickr.com._

I would like to thank the photographers who have given permission for their images to be included. Credits for these are provided within individual photograph captions and where no name is given, the photographer is myself, Andy Chard.

Facts have been checked and cross-referenced as much as possible and if any readers are aware of any errors or useful supplementary details, the author would be pleased to receive these via the publisher's address on the title page or by email to updates@platform5.com.

INTRODUCTION TO
LOCO BASHING

As various members of the class had done many times before, on Saturday 12 March 1988, 31418 came to a standstill at Stockport station while working the 14.22 Sheffield–Liverpool. "Excellent, I need this one" I proceeded to tell the handful of bashers I was with, as we took seats in the Mark 2 open coach. It would be my 64th Class 31/4 haulage, which aside from the two ETH-equipped sub-class members which had already been withdrawn, would leave me needing just four more to track down for haulage. As the driver applied power and the coaches edged forward, the engine abruptly shut down, bringing the train to a standstill five metres forward of the position we had boarded. At that point we didn't know the train would be terminated and the hope of a rare rescue loco hadn't yet been extinguished. A discussion then began about whether we'd had 31418 for haulage or not and what constitutes being hauled by a loco, which after all, was the object of our hobby.

It was an interesting conversation, raising the wider question of what exactly is bashing – this specific, yet quite diverse variant on the spectrum of being interested in trains and railways. Bashers enjoy travelling on trains, steadily accruing haulages and mileages, most commonly but not exclusively, from diesel locomotives. The sight and sound of these powerful beasts at work, as they convey the basher to various corners of the country, is as much the attraction, as the challenge of working towards clearing a class for haulage – a significantly greater achievement than simply seeing them all.

The consensus that afternoon in Stockport was that those five metres along Platform 3 were insufficient and that a loco haulage needed to be from one station to another. I loosely agreed, but like anything in life, there are grey areas and exceptions. I speculated that if after having had a loco from Crewe for example and it embarrassed itself as 31418 had

> **Below:** Having put the previous year's incident which generated a lively discussion about loco haulage behind it, 31418 arrives at Ely with the very amply loaded 09.16 Cambridge–Kings Lynn on Monday 8 May 1989. *Graham Walker*

done, half a mile short of Chester, surely having travelled more than 20 miles behind it was sufficient to warrant a haulage, despite not qualifying for the station to station definition. And what about shunting locomotives, which don't work timetabled services between stations, but provide haulages when they move portions of sleeper trains with passengers on board, would that qualify?

At that time, preserved loco haulages on heritage railways or those arranged in advance to work railtours were generally seen as cheating, or of secondary value to those on main line service trains. Today, however, such distinctions are rarely, if ever made, as heritage railways and main line tours are the only way to experience haulage behind many classes. A number of fledgling heritage railways don't even have two stations for trains to travel between, which further challenges our 1988 definition!

As the opportunities for main line loco haulage have narrowed over the years, the question of what constitutes a haulage has widened. What happens, for example, when a loco is converted to something new – if one had travelled behind it in its former guise, is it required under the new identity? After a series of Class 47s were re-engined and rebuilt

around the turn of the century, they were reclassified as Class 57s. Few would claim that haulages from whatever remnant of the former Class 47 that survived in the new loco constituted a haulage from the new Class 57. Similarly, the six Class 37/9s that received modifications and entirely different engines between 1986 and 1987, were still classed as 37s, but many believed the changes were significant enough to consider these as new locos, creating six new members of the class to track down for haulage.

There are more shades of grey for locos with two forms of power. Electro-diesel Class 73s for example can operate on diesel or electric power and when working on the Southern Region, they would sometimes change between diesel and electric power while hauling a train. Most bashers considered either form of power sufficient for haulage, although some would note down or aim for instances of each. Similarly, Class 88s can haul trains in either diesel or electric mode and Class 92s can draw power from both AC overhead electric and DC third rail sources.

All of this illustrates that there any many different ways to enjoy the hobby and fortunately there is no

Below: Westerns and Deltics are two of the classes credited with accelerating interest in diesel bashing as they gained popularity and headed towards cult status through the 1970s. They were also pioneers in the diesel preservation movement as this view of D1041 shows, when WESTERN PRINCE departs from Haworth on 5 November 1988, during a very early diesel gala at the Keighley & Worth Valley Railway.

Above: Multiple units could never match the attraction of diesel locomotive haulage, but in the course of chasing diesels, a considerable number of journeys involved EMUs like this. During the 1980s, Class 304s worked trains from Altrincham to Manchester's Deansgate, Oxford Road and Piccadilly stations every 15 minutes, with direct services continuing to Hazel Grove, Alderley Edge and Crewe. 304008 heads for Altrincham on 24 January 1987. *Richard Allen*

governing organisation or regulatory committee issuing edicts as to what is and isn't valid. We can therefore each create our own rules and targets and find satisfaction in partly or fully achieving these, without having to look over our shoulder at what someone else is doing.

The answers to the question of what loco haulage consists of are almost as diverse as the number of variations of the hobby. In the same way that music has genres within genres, bashing has various sub-cultures. There are those only interested in haulage from one class or engine type (Sulzer or English Electric for example), with some phenomenal mileage figures or numbers of haulages from the larger classes having been achieved; these are explored further in Appendix 3. Some bashers have pursued as many loco haulages as possible across various locomotive classes, irrespective of mileage. Others will do a bit of everything, accumulating a variety of haulages, concentrating on favourite classes or mileage at times, with a bit of line bashing

and photography thrown in too. When a class has been cleared for haulage, some record different combinations of loco pairings, or the examples that have hauled them along a favourite line. Then there are the enthusiastic adherents who reset the clock on the first day of January and start the process again each year. This book focuses on locomotive haulage, however I have come across some who bash units and I used to know a chap from South Manchester for example who loved Class 304 EMUs and would regularly travel on these before and after work to increase his mileages. Only recently (summer 2019), I came across a new strand of bashing while chatting to someone at Doncaster station who recorded which HST power cars he'd had for both pushing and pulling and he was impressively close to clearing the lot for both. There will no doubt be other variations of the hobby that I haven't yet encountered.

The recording of loco haulages dates back to the days of main line steam. As the number of steam engines built was significantly greater than that of diesels (tens of thousands, as opposed to thousands of diesels), the challenge of sighting as many as possible was sufficient to occupy most steam enthusiasts. There were, however, those who also noted which steam engines they had travelled behind, some of whom specifically pursued steam haulage. Steam bashers were few in number, but they were the pioneers of haulage seeking nonetheless. After steam was eliminated from the main line in 1968, large numbers of the generation that had grown up spotting steam engines drifted away from the hobby, or directed their attention to the emerging heritage railway movement. Among the next generation of enthusiasts, increasing numbers only knew diesels and certain locomotive types began to grow in popularity. These attracted a following, many of whom would seek to prolong their encounters by travelling behind the locos, with Westerns (Class 52s) and Deltics (55s) having significant fanbases from an early stage.

The incentive to make the most of travelling behind these popular classes increased in the early 1970s, as it became clear that the days of the hydraulic classes were numbered; Classes 41–43 had been eliminated by 1972 and large numbers of Hymeks (35s) and Westerns had also been withdrawn by then. The arrival of production HSTs (which later became the other Class 43s) and Class 50s from the London Midland Region in the mid-1970s signalled the end for the remaining Westerns; the Deltics were already heading for the same fate. This cemented the cult status of these classes among their followers, who continued to make the most of their workings. At the same time, classes which still had many years ahead of them were growing in popularity and these included the 25s, 37s, 40s, 45s, 47s and 50s, all of which had their own haulage aficionados. The variety of diesel types, destinations to travel behind them to and the pursuit of clearing

Below: Electric locomotives were never as high a priority as diesels, but being hauled by any new loco is always welcome. Class 90 No.90026 is just one week into its working life on 1 April 1989. It is seen at Stafford in revised version of the InterCity livery with the 17.05 London Euston–Holyhead, complete with a sleeper coach at the front of the train.

Above: By the 1980s, Class 40s had built up a strong following, before their last regular passenger workings took place in 1984. There is evidence of this on 24 July 1982 when 40006 took charge of the 09.00 York–Llandudno, seen to the west of Colwyn Bay during the latter stages of this working. *Gavin Morrison*

classes all fuelled the popularity of diesel bashing. Some were drawn into it by existing practitioners and others made the progression from recording sightings to a general pursuit of loco haulage. As the 1970s turned into the 1980s, bashing remained popular, although it suffered setbacks when classes with an established entourage in tow were gradually all withdrawn, including the Rats (25s), Whistlers (40s) and Peaks (45s).

For me bashing is as much an experience as it is a ticking off exercise. Observing the varying geography of the British counties through the carriage window, exploring destinations that I would never otherwise have reached and trying to find a fish and chip shop in these are all part of the adventure. I enjoy photography and the challenge of taking a decent picture, which I am still learning the art of today. The sound of the thrash is of course a large part of the appeal and why I prefer to travel behind locos, over observing them at stations or from the lineside.

The type of coaching stock makes a big difference, my preference being Mark 1 or the early Mark 2 variety, with their opening internal windows, which allow the traction to be heard. Most loco-hauled stock has a superior internal layout to 21st century built vehicles, which unlike their predecessors, tend to lack tables, legroom and seats that line up with windows. The earlier designs also facilitate striking up conversations with other bashers and socialising around the tables.

Having engaged with this quirky hobby for over three decades, it's been interesting to observe its culture and behaviours and how some of these have changed. In the 1980s, it was quite common for children to travel unaccompanied to another region, as I did from the age of 12, making solo trips to cities across the North of England with my parents' full confidence. I wasn't allowed to stay out into the evening until a few years later, or do an "overnight" until I was 16, but there were plenty of others with less stringent requirements, which would be seriously frowned upon today.

Various terminology has evolved among bashers and much of the dialect that was widely spoken when I joined the cause survives today. Words still widely used include "Ned", a light-hearted acronym for new engine desperado, describing junior bashing practitioners and "clag", being black engine emissions. Others that I haven't heard for some time include referring to an engine as "dreadful"

9

as a term of appreciation, in contrast to the word's orthodox meaning and "normals" when discussing people who aren't enthusiasts! The pages that follow include a number of these words, the meanings of which are explained in Appendix 1 for the benefit of the uninitiated.

Before the 1990s, all trains had windows that opened on every carriage door, with further opening windows on the inside of most carriages and multiple units. Whilst there were stickers warning of the dangers of doing so, leaning of out of windows was widely practiced. Bashers would usually sit in the front coach of diesel hauled trains, to enjoy the sound of the loco(s) working hard. At stations, they would often poke their heads out of the window as the train departed, for a better view and maximum audible pleasure. At times and usually at lower speeds, some would also extend their arms laterally with the hand pointing upwards, a much higher risk practice known as bellowing or flailing, to

indicate their appreciation of the traction. Leaning out of carriage windows and the waving of one's arms was not exclusive to bashers though. Plenty of people would be seen leaning out of windows, as trains arrived and departed from stations, some waving at friends and family when meeting arrivals or seeing people off from the platform, which was much more commonly practiced. Similarly, if a train stopped between stations, or was travelling along a scenic section, heads would appear from windows, many of whom hadn't a clue what the loco on the front was. When manually-opened slam-door stock was standard, it would also be common to witness commuters opening train doors and alighting before trains had stopped! That was well over a generation ago though and things are different now, with the dangers of leaning out of train windows sadly being illustrated by two fatal incidents recently. The ubiquity of phones with cameras and internet capability also means that behaviours which were

Above: D7612 (25901) waits to attach to the rear of the stock as D7535 (25185) brings the 12.53 from Minehead into Bishops Lydeard (both locos were on loan from the South Devon Railway) on 21 June 2019. At the end of the day, a handful of diesel devotees who had never met before and may never meet again, shared a very sociable journey on the last bus from Bishops Lydeard to Taunton. This is typical of the friendly nature of those who enjoy the hobby.

previously only witnessed by those present at the scene, can now be made public and permanent within seconds, jeopardising the future of the basher's preferred locomotives and coaching stock.

Bashing attracts all sorts of interesting and colourful characters, with some becoming well known for where they can be found or their endearing eccentricities. The vast majority are knowledgeable, sociable and very helpful people. I often end up chatting to fellow enthusiasts and have had countless conversations with those I've not seen before or indeed again, sharing workings information and various bashing tales. In the months before writing, these have included a spontaneous delegation on the last bus from Bishops Lydeard to Taunton after the West Somerset Railway's 2019 Diesel Gala and a fascinating conversation with the security staff while I was photographing 37418, after travelling behind it from Cardiff to Rhymney. I learned not only about some of the disappointing behaviours that my fellow enthusiasts exhibit, but also of an unexpected, newly found affection for Class 37s that this non-enthusiast employee had developed, perhaps just becoming their newest fan!

The expectations of many bashers have changed over the years. When I started, the vast majority of bashers simply went out on a given day without any knowledge of where particular locos were, to simply see what could be found that day. Some gen was passed on verbally, but this was ad-hoc and generally only concerned workings for that particular day. Diagrams were available, with publications such as Loco-Hauled Travel, however this only provided an indication of which class or sub-class to expect on particular trains, with varying degrees of accuracy. Today, main line diagrams are covered by very small pools of locos and the times and dates that individual locos will work on heritage railways are available in advance. Railtours also advertise the traction (although not often the specific locos) in advance and all of this together has created the expectation to know what is on offer beforehand, with some people deciding whether to travel based on this information. It is very helpful knowing what to expect, but this is at the expense of much of the fun of pleasant surprises or disappointments that used to characterise the hobby. These also increased the satisfaction when one would unexpectedly come across a rare or required loco.

There are practically unlimited historic locomotive workings, diagrams and moves that could be documented in a book about bashing, so this book is primarily an account of my exploits and unique experiences while pursuing diesel haulage. These have been enhanced with plenty of contextual and historical information on the locomotives themselves, workings and railway history. As it is a personal account, most of the photographs used are my own, to illustrate the tales that the words tell. The words have also been complemented with some brief details from my life, to further ground the material. I have tried to take the reader back, to see through the eyes of a young basher in the 1980s, from when I first became interested in railways, to becoming a fully-fledged basher. It was a formative time and yet even as a teenager, I knew that locomotive types, railway infrastructure and practices which originated in the Victorian days were disappearing. Increasingly few pockets of these remain on the main line today, although the heritage railway movement has done a wonderful job of preserving many of them. I hope that the pages to follow are another small way in which details of what were once everyday sights and sounds, and one strand of rail enthusiasm, can be documented.

Andy Chard
Manchester, October 2019

Below: 47537 is seen at York on 30 May 1988 while working the 09.24 Bristol–York, as part of a diagram which next saw it work an unusual train - the 16.11 York–Liverpool Lime Street which ran via Birmingham New Street! The 47 therefore returned to Leeds that afternoon, hauling the circuitous train to Liverpool as far as Birmingham. It is carrying the name "Sir Gwynedd/County of Gwynedd", which it did for a decade after being named at Holyhead station in 1982.

CHAPTER 1 –
SETTING THE SCENE

I'm not sure where it came from, but I have always been interested in railways and love travelling by train. It's not only the trains that interest me, but the routes the lines take, the variety of railway architecture, the icons that are semaphore signals and the atmosphere that these all create. I grew up in Sale, a suburb of south Manchester, which before being closed for conversion to the first section of the Metrolink (light rail) network, was 5.75 miles from Manchester Piccadilly on the Mid-Cheshire line to Altrincham and Chester. During my childhood, my Mum would often take me and my two younger brothers on the 2.75 mile journey to visit the shops in Altrincham, where the electric trains terminated. Hourly Chester trains, which started from the bay platform at Manchester Oxford Road, would stop at Sale and Altrincham only, before continuing into rural Cheshire, providing non-stop services in either direction from Sale. During school holidays the four of us would take a Class 101 or 108 diesel multiple unit (DMU) directly to Delamere to visit the forest there, or continue to Chester or Llandudno for a day trip, while my Dad was at work in Manchester. I loved these trips and for me, the train journeys were at least as exciting as the destination. From the age of 9 or 10, my parents would put me on a long first generation DMU at Manchester Victoria and ask the guard to keep an eye on me during the journey to Blackpool North, where I would be met by my grandparents. Many a happy weekend or school holiday was passed this way and I loved those journeys and the sense of freedom when I was travelling by train, especially without adult supervision. Noticing my interest around this time, my Dad bought me a copy of Ian Allan's 1981 ABC Locomotives book and suggested I start recording the numbers of the locomotives I saw. And so began the first form of my as yet, unending locomotive caper, which would evolve and specialise in time.

I was a very independent child, disappearing off to the park or to local hangouts on my bike, such as the banks of the River Mersey, which is no small watercourse as far upstream as south Manchester. After I started secondary school in 1984 at the long-since demolished Sale Boys' Grammar School, my appetite for further freedom grew. Wanting a closer acquaintance with the locomotives represented by the numbers in my Ian Allan book, I made my first solo outings to Manchester sometime during 1985, at the age of 12. It was easy enough to take the train from Sale to Manchester Piccadilly, or the bus down the A56 to reach Victoria station. Child bus fares were a flat 10p charge anywhere in Greater Manchester and platform tickets cost 6p, making a very economical afternoon's entertainment. Between the two stations, I found plenty of examples of Class 25s, 31s, 45s, 47s, various electric locos and the occasional HST. There were also many diesel and electric multiple units, but these were neither in the little book which my Dad had given me, or as inspiring as the more powerful and noisier locomotives I found, many of which had intriguing names.

I found out that Dave Goodman, a lad in the year above me at school who I vaguely knew, was more experienced and well-travelled in the hobby. Our mothers knew each other and they brokered a deal for me to join Dave and his friend on one of their planned trips, to what for me was the exotic destination of Crewe. So on a distinctly cold Saturday 2nd November 1985, as we sped through Cheshire behind BR blue liveried 86219 "Phoenix", Dave asked me if I needed it for haulage. I had to ask him what that meant and he explained that as well as recording which locos you have seen, you can also score them for haulage once you have travelled behind the loco and that doing haulages was apparently pretty popular. "In for a penny, in for a pound" I thought and from that day I wrote a letter H after each engine that I had for haulage.

As 1985 turned into 1986, I started to explore further afield and made solo trips to Sheffield, Chester and Crewe a number of times. There were always plenty of enthusiasts at stations in those days who I would strike up conversations with and gradually I increased my knowledge of trains and railways. Initially I primarily pursued locomotives for sight and made good progress with this, while continuing to note every loco haulage I had. I was out and about during most weekends and school holidays and in mid-1986, about a year since I started, I began to make regular returns from Manchester Piccadilly to Stockport and from Victoria to Stalybridge and Bolton. This could be done with a Peak Wayfarer ticket, which provided unlimited travel on trains

Two photos of the innocuous and long-gone mode of transport with which many a day's bashing would begin and end. Before Manchester's Metrolink revolution, hourly DMUs between Chester and Manchester travelled through Sale, running non-stop to the city centre and Altrincham in far less time than the Metrolink takes today.

Top: With Northwich station just visible beyond the bridge, a Class 108 on one of these hourly services heads for Chester on 1 July 1988.

Bottom: A Class 108 pauses at Sale with the 16.25 Manchester Piccadilly–Chester on 15 February 1987. *Richard Allen*

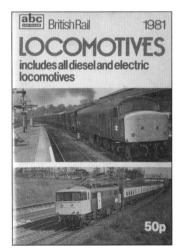

Left: We all start somewhere and with an empty book. This is the first one that I began to annotate.

Below: The scene that typically met me at Manchester Victoria at the time; 45150, 45034 and 47203 are stabled in the centre roads, another 47 and a 31 rest in the distant eastern sidings and further diesel locos are likely to be stabled in the bay platforms where the Metrolink station is now situated. 142035 is about to head east and when this photograph was taken on 9 May 1987, the Pacer was at the start of its life, heralding the beginning of the end for many loco-hauled trains. At the time of writing in late 2019, roles have reversed and Pacers now find themselves on borrowed time and diesel loco-hauled trans-Pennine services are returning to Manchester Victoria. *Richard Allen*

and buses in Greater Manchester and beyond for a day. For me, travelling by train has always been an inherent part of the interest and as well as seeing more of the rail network and more locomotives, I gradually notched up more haulages.

There were plenty of locomotive haulages on offer in Greater Manchester in 1986, within the validity of a Peak Wayfarer ticket. At Piccadilly, Class 31s worked the hourly services to Sheffield and beyond and there was a huge variety of Class 47 hauled trains on direct services to and from London Paddington, Harwich,

Brighton, Newhaven, Cardiff, Swansea, Portsmouth, Poole and Paignton. The AC electric Classes 81, 85, 86 and 87 were in charge of what in those days were hourly trains to and from London Euston and Birmingham International, plus the northern leg of the direct trains to Dover Western Docks, Plymouth, Newquay and Penzance. From the May 1986 timetable change, the newly constructed Hazel Grove Chord created a link between Hazel Grove on the Buxton line and the freight only line which passed above this. The new link was less than 500 metres

Above: ETH-fitted Class 31/4s on short rakes of Mark 2 coaches were "two a penny" in Manchester during the mid to late 1980s, working the hourly trains to and from Sheffield, before being displaced by Sprinters in May 1988. 31454 approaches Edale on the 10.41 Manchester Piccadilly–Sheffield on 8 June 1985. *Richard Allen*

long, but it brought new opportunities for the haulage seeker operating on a budget. Instead of being routed via New Mills, the hourly services from Manchester to Sheffield now travelled via Stockport. This, together with the many Class 47 and electric-hauled long distance services, provided a vast variety of traction between Piccadilly and Stockport, Wilmslow or Macclesfield. Secondly, instead of starting from the bay platforms at Piccadilly, the trans-Pennine trains between Manchester and Sheffield now continued to Liverpool. The longer journeys, together with the fact that many of the trains also continued to Hull, Cleethorpes or Great Yarmouth, required a greater number of Class 31s to be in operation each day. There was a comfortable hourly move available, taking the 31 from Manchester Oxford Road at 34 minutes past the hour (or 38 minutes past from Piccadilly) to Stockport, where they arrived at 47 minutes past. The next Class 31 left at 10 past the hour, arriving into Oxford Road at 23 past, which was 11 minutes before the next 31 was due.

I enjoyed the great variety of diesel engines that could be found at Victoria station. Locos passed through on freight trains and could also be found stabled in the sidings, centre road and former bay platforms which are now occupied by the Metrolink platforms. Two Class 45s would pass through every hour on the Newcastle or Scarborough to Liverpool, Bangor and Holyhead services. These were timed well to allow an hourly 15 mile return trip to Stalybridge, giving a manageable 9 minute connection there and 21 minutes at Victoria before the next Peak up Miles Platting Bank. There were also long distance loco-hauled trains between Manchester Victoria and Bolton, such as the morning Manchester and Liverpool to Glasgow and Edinburgh service, which had Class 47 hauled portions from Victoria and Lime Street stations. These were joined at Preston; an AC electric then took the train forward to Carstairs, where they were separated into portions again for Glasgow and Edinburgh. Curiously, there was no southbound return working for this train, except on Sundays. There were also three north-south, long distance services which crossed Manchester. As this was before the Windsor Link (linking Deansgate and Salford Crescent stations) had been built, through trains used the longer and more interesting route from Stockport, through Denton, Ashton Moss North Junction and Miles Platting. These trains were usually in the hands of Class 47s and Table 1 summarises the regular workings.

Table 1: Loco-hauled trains running north / south through Manchester Victoria Monday–Saturdays during the Summer 1986 Timetable (12 May–28 September).

Diagram	Northbound Services	Stockport	Victoria	Bolton
1	07.50 Manchester / 07.45 Liverpool–Glasgow & Edinburgh	n/a	0750	0806
2	07.15 Nottingham–Glasgow & Edinburgh	n/a	0959	1023
3	07.20 Harwich Parkeston Quay–Glasgow & Edinburgh	1304	1335	1350
4	13.20 Harwich Parkeston Quay–Blackpool North	1856	1920	1935

Diagram	Southbound Services	Bolton	Victoria	Stockport
3	07.20 Blackpool North–Harwich Parkeston Quay	0820	0843	0906
4	10.15 Edinburgh / 1020 Glasgow–Harwich Parkeston Quay	1352	1415	1438
1	15.15 Edinburgh / 1520 Glasgow–Nottingham	1903	1923	n/a

As it's a good 20 minute walk between Piccadilly and Victoria stations and Manchester's weather doesn't have the best of reputations, the loco hauled trains between Stockport and Victoria were a useful way of avoiding the walk. Another alternative was to take one of the "Table 92" services between Stockport and Stalybridge, via Guide Bridge where there would

Below: BR named the Harwich to Glasgow & Edinburgh train "The European" and this was one of the few daily services to criss-cross its way across Manchester via Denton and Ashton Moss Junction. On 17 September 1986, 45013 heads into Buxworth Cutting, near Chinley, on the 10.20 Glasgow Central–Harwich Parkeston Quay. *Neil Harvey*

Below: Making a similar journey to that which 47464 made a couple of years earlier, 47653 departs from Shipley with the 14.20 Leeds–Carlisle on 29 May 1988. As well as the trains on the route, this location has changed beyond recognition, with the signal box and semaphores long gone and a mesh of overhead catenary now blocking the open skies.

usually be a good selection of diesel locos stabled between freight duties. In the late 1980s, a first generation DMU would make an hourly return trip between the two Cheshire towns from early morning until late evening, unlike today where there is a sole weekly "ghost train" on this route.

As soon as I turned 13 in 1986, I started a paper round and the proceeds from this provided the funds for more regular rail exploration. It was a weekly delivery of the local free paper and I had the flexibility of doing it any time each Wednesday to Friday and would often do additional rounds to cover for other people and earn a few extra quid. I generally didn't have the means to travel further than about 50 miles from Manchester; however, when opportunities for free long distance trips came, I would jump at the chance. My grandparents lived in Blackpool and I would often stay with them for weekends, or longer during school holidays. Around this time they discovered the benefits of a Senior Citizen Railcard; they received a third off all rail fares

and accompanying children could travel anywhere in the country for just £1. I may have planted the idea for some of the many trips that followed, the first of which were on two consecutive days in May, followed by a day out to York a few weeks later. As Table 2 shows, we didn't do too badly for loco haulage on our first of these outings, with haulages from Classes 31, 43, 45, 47, 81, 85 & 86 and a scenic loco-hauled round trip on the Settle & Carlisle and Bentham lines.

Another new avenue of rail exploration unexpectedly opened up in 1986. During a family day out to explore the Keighley & Worth Valley Railway (KWVR) and the attractions of Haworth three years earlier, my parents stumbled across an intriguingly cheap dilapidated small house for sale. To say it was in need of improvement was an understatement, as it had no electricity, hot water, inside toilet or bathroom! Not shy of a challenge, they had used a recent inheritance to buy the bargain property, which my Dad then spent three years labouring to renovate, providing us with a new base for many a weekend break and family

Table 2: Moves with Grandparents during mid-1986

Date	Haulage	From	To
28 May	DMU	Blackpool North	Preston
28 May	86435	Preston	Carstairs
28 May	47704	Carstairs	Edinburgh
28 May	47468	Edinburgh	Carstairs
28 May	85031	Carstairs	Preston
28 May	31453	Preston	Blackpool North
29 May	DMU	Blackpool North	Preston
29 May	85014	Preston	Carlisle
29 May	47646	Carlisle	Skipton (via Settle)
29 May	47464	Skipton	Lancaster (via Bentham)
29 May	81007	Lancaster	Preston
29 May	DMU	Preston	Blackpool North
7 July	DMU	Blackpool North	Manchester Victoria
7 July	47381	Manchester Victoria	York
7 July	47445	York	Manchester Victoria
7 July	DMU	Manchester Victoria	Blackpool North
20 Sept	47556	Manchester Victoria	York
20 Sept	43069/195	York	Durham
20 Sept	45130	Durham	Manchester Victoria

Above: Between May 1985 and September 1986 a variety of Class 33s visited Manchester while working the 07.50 from Swansea and 13.45 return working to Cardiff Central. The loco would run round by way of the crossover which is visible towards the buffer stops. On Monday 13 May 1985 33043 was in charge of the first of these workings, when there was not a multiple unit in sight and Platform 1 hosts a rake of Mark 2 coaches which will be the Class 31 hauled 13.41 to Hull. *Richard Allen*

holiday. Not only did the large living room window look down on a stretch of the heritage railway between Haworth and Oakworth where steam trains would pass as we ate, but this provided me with a second base from which I could regularly explore the railways on the eastern side of the Pennines. I had the choice of taking a steam engine or a Waggon & Maschinenbau diesel railbus on the KWVR, or the bus to Keighley. From there a DMU on a stopping service or a loco-hauled train from Carlisle could take me to Leeds and beyond. There were only three services per day in each direction between Leeds and Carlisle at the time and with haulages gradually becoming more important to me, I tried to time the Keighley to Leeds legs of my travels to coincide with these as much as possible. One such occasion was Friday 15 August 1986, when I was delighted to find BR green-liveried 40122 bring a lengthy rake of Mark 1 coaches into Leeds on the 13.00 from Carlisle. After running round the stock, I took the loco to Keighley on the 16.05 return working to Carlisle. That was the only occasion I had a Class 40 on a service train, having just missed the last of the Class's regular workings in 1984, shortly before I took up the hobby. Having had a number of 40s since, both on heritage railways and main line tours, I'm confident that had I been born a few years earlier, I would have been a keen Whistler basher. Then again if I grew up in the South or Midlands during a different era when alternative traction was more accessible, perhaps 40s wouldn't have had a look in.

A week or so later when I was back in Manchester continuing to make the most of the school holidays, I had a couple of unexpected winning haulages. On 26 August, the very unfamiliar looking and sounding 33044 rolled into Stockport on the 07.50 Swansea–Manchester, which I gladly boarded for its final six miles. I didn't know at the time that this was a Monday to Friday only Class 33 diagram, which I hadn't come across before with being at school on most weekdays. The Crompton engine would run round using the crossover at Piccadilly station and return on the 13.45 Manchester–Cardiff Central. I was back for more three days later and I doubled my Class 33 haulage tally courtesy of 33033 on the same working. Had I known that this service would finish a few weeks later and there would never be another Class 33 diagram to Manchester, I would

have made sure I did it a few more times. There were, however, other activities planned, some of which I didn't have control over and the following week our family were staying in Haworth again. One of the dates that my meticulous record keeping has failed me was somewhere around 3 September 1986, when a family day trip to Scarborough proved fruitful for Class 45 haulage, especially given that my parents and not myself would determine which trains we travelled on. I may not have recorded the date, but I certainly didn't fail to note which diesel engines hauled us across Yorkshire; after taking a DMU from Keighley to Leeds, 45128 took us forward York, which was presumably on a Liverpool–Newcastle service, necessitating the change of trains. 45133 then transported us to Scarborough, with 45106 and 45116 returning us on the corresponding return moves.

Outside of the routines which school and family life dictated, the latter months of 1986 consisted of a number of short and medium term ventures from one of my three bases; home in south Manchester, our small holiday home in Haworth and my fellow train-loving grandparent's house in Blackpool. I continued to frequent the rail network most weeks, steadily adding to my haulage tally largely in the form of further members of Classes 31, 45, 47, 86 and 87 on their many passenger services. The following year however would turn out to be one with plenty of new destinations and interesting workings to be discovered, both locally and further afield.

CHAPTER 2 –
BECOMING A BASHER

The nature of my interest started to change sometime within 1987 and looking back, I can see the factors that contributed to this. At the beginning of the year, I started to realise that finding new locos for sight was becoming difficult after less than two years at it, as my success rate was down to around 10% on each outing. Even on long-distance journeys to new destinations it was only a little better at about 20%. I enjoyed travelling more than the static variant of the hobby and peacefully watching the country sail by, through a train window remains one of my favourite experiences today. I also saw that acquiring new haulages provided a much more interesting and diverse challenge. In addition, I have always found diesel locomotives the most compelling feature of railways. I vividly remember walking over the open air footbridge between Platforms 5 and 6 at Crewe around early 1987, to go and examine one of the recently converted ETH-fitted Class 37/4s in Large Logo livery (37426–37431) that had just arrived in one of the bay platforms. I nearly jumped out of my skin when the driver applied full power as I was practically above the engine and I thought something along the lines of "what a complete beast of a machine that is". Perhaps that was when I first began to appreciate the sound of diesel thrash, which is another of the ingredients that directed the evolution of my hobby.

I continued chipping away at new haulages, with there being plenty to go on in the North-West. The number of locos I'd had for haulage in early 1987 could be counted in dozens, leaving vast quantities and types of engines untapped. Being out travelling increasingly regularly meant that I got to know some of the regular characters from the region, who were a friendly and colourful bunch. Through them I learned about NBs, as they were known and how these were much bigger haulage trophies. NB stood for "Non-Boilered", meaning the diesel loco had either not been built with a steam heating boiler, or its boiler had been isolated or removed. NBs

Below: The sunlight suggests that this view was not in the middle of winter. Train heating was definitely required, though, when 47483 crossed Ribblehead Viaduct with the 07.55 Dundee–Poole on 21 February 1987, ending the misguided hope that the diverted West Coast Main Line trains might involve some no heat traction. *The Michael McNicholas Collection*

Above: In the late 1980s, Class 50s storming past each other on the Great Western Main Line was an everyday occurrence. As Network SouthEast livery 50030 approaches a very different looking Slough to that of today with the 14.41 Birmingham–Paddington, the year after it became my first Class 50 haulage, 50012 heads west with the 16.52 Paddington–Newbury on 25 August 1988.

differed from other classes and sub-classes which had electric train heating and air conditioning equipment fitted; ETH-fitted locos included Classes 31/4, 47/4 and 50, which were routinely, but not exclusively, found working passenger trains. In the 1980s, British Rail had assigned locomotives to one of five different sectors and these were InterCity, Provincial, Network SouthEast, Railfreight and Parcels. NBs were mostly freight sector locomotives and included large numbers of Classes 31/1, 37/0, 47/0 and 47/3, which rarely worked passenger services. In the warmer months, however, when heating wasn't needed (especially on Class 2 trains where air conditioning consisted of sliding the internal windows open), NBs could be found on passenger trains. One could simply be in the right place at the right time and find a NB Class 31, 37 or 47 on a passenger train; however, in time I would learn that they were more common on the summer Saturday trains to holiday destinations such as Great

Yarmouth, Skegness and Aberystwyth. An NB was a big winner, as the loco may not have worked a passenger train for many years and it could be years, if ever, before it would do so again.

One of my first attempts to track down some NBs was on Saturday 21 February 1987. I had learned that the West Coast Main Line would be closed between Preston and Carlisle due to engineering works and naïvely thought that BR would be throwing out freight locos on InterCity services, which were being diverted via Blackburn, Hellifield and Settle. My plan was to travel to Preston, buy a return to Blackburn and see what turned up. With hindsight, there was no way that BR was going to put "no heat" locos on trains that would take well over two hours to meander their way between Preston and Carlisle across one of the most exposed areas in the country in February. I was therefore a little disappointed as I found the likes of ETH-equipped 47465, 47628 and 47483 doing the honours.

A few weeks later my Mum had to travel to London and offered me and a friend (who never did catch onto the hobby) the opportunity to join her and explore a part of the South-East of my choosing. I remember discussing with school friends which loco classes might be found at which London stations, but in truth we knew very little about what could be found at each. At the time, Class 50s were the only main line diesel class that I hadn't seen any examples of, let alone had for haulage and the consensus was that Paddington was our best bet to find some. And so on Saturday 21 March 1987 after 86405 took us to London Euston and we crossed the capital via the Circle Line, I will never forget the moment I walked up the stairs from Paddington's Underground station to the main concourse. The hope that I would track down a member of this exotic class was rewarded with the imposing sight of Large Logo liveried 50012 "Benbow" on the blocks at Brunel's grand terminus. By the time I had bought a return ticket to Reading, 50012 had repositioned itself at the front of the next train west, so I made my way to the front compartment on the rake of Mark 1 coaches. As "Benbow" powered away from Paddington station, providing me with my inaugural 50 haulage, my captivation with the class was confirmed. I was never going to be able to clear the class for haulage, as 50011 had already been withdrawn and 50006 & 50014 soon followed, but it was something that I would have a good crack at over the next few years.

A number of the lads at school were also into trains and some would join me on my travels, one of whom was Paul Gerrard, who also lived in Sale and was in the same year as me. There was also a contingent who would regularly go plane spotting at Manchester Airport, noting the numbers of the civil and military aircraft they saw in the aviation equivalent of Platform 5 Pocket Books. Sometimes a "rare working" would cause some excitement in the same way as an NB would on the railways and I remember wondering if there was such a thing as a plane basher and how difficult that would be. Whilst the two factions had many similarities, they tended to peacefully co-exist rather than overlap and I was never tempted to dip my toe in that particular water.

Below: 45128 provided some drama and the hope of a rare rescue engine when it set on fire near York on 21 April 1987. Having put that matter behind it, the loco performed more reliably on 15 February 1988, when it is seen at Chesterfield in the afternoon winter sun while working the 13.25 Nottingham–Leeds.

Above: A few colourful examples from one of my sideline interests – collecting destination labels, especially the more coveted unused examples, before they were applied to carriage droplight windows. I would write the service details and loco that was in charge on some that I collected, as I have done for the Network SouthEast label, as a memento of the train, which in that instance was hauled by 50035.

Engine failures were an opportunity for interesting and unusual loco haulages. The sighing and negative remarks from the general public contrasted sharply with the excitement among any enthusiasts on board, who would speculate about what traction the local BR control centre might scramble to rescue the stranded train. One could end up being dragged by something very rare, such as a pair of 20s that hadn't worked a passenger train for many years, or a 56 that may never have done so before. Sometimes, the rescue engine(s) would only haul the coaches forward to the next major station, where the train would then terminate, meaning that if you were on it when it failed, you could hit the jackpot. The unexpected and potential rarity value was part of the variety and fun of bashing. So on 21 April 1987 when I was happily riding behind 45128 on a TransPennine service with my parents and brothers for one of our family days out to York, when our train ground to a halt a few miles south of York, there were some very promising signs. I looked out of the droplight window to see orange flames leaping out of the Sulzer engine's side grill and doubted it would be taking us any further that day. Half an hour later, a fire engine was fighting its way through the crops of the field adjacent to the railway in order to deal with the naughty Peak. Our driver had put detonators on the four track section to warn passing trains of the danger with loud explosions, adding to the drama of the occasion. As far as I could see, I was the only enthusiast on board and when we were finally rescued, it was nothing close to the rare freight traction I was hoping would be sent from York, but the comparatively everyday 45111, leaving me feeling a little bit cheated.

My next on-board locomotive failure and rare haulage rescue opportunity came just four days later. Returning home from another day out on the other side of the Pennines, I was heading west

Above: A typical example of a rare rescue engine. On 20 March 1986, 45116 failed near Marsden while working the 07.10 Newcastle–Liverpool; Eastfield-based 37014 was sent from nearby Healey Mills depot to rescue the train, taking it forward to Manchester Victoria where it was terminated. The no-heat loco then ran round and formed the 12.05 Liverpool–Newcastle, which started in Manchester. It is seen after arriving at Leeds, where it was replaced by 45134 which took the train to Newcastle. *Simon Record*

along the Hope Valley Line behind 31466 on a late afternoon Sheffield to Liverpool service. It was clearly struggling with its short load of four and took nearly half an hour to limp through the 3.5 mile-long Totley Tunnel. The signaller diverted us into Earles Sidings at Hope, which cued excited speculation from the fellow enthusiasts on board about what was working the Hope Cement trains that could take us and our errant Type Two forward. For the second time in a week, I felt like more of a loser than a winner, as it wasn't long before 31412 on the next hourly Class 31 hauled Sheffield to Liverpool train also swung into the sidings. The only rare workings we were treated to that day were traversing Earles Sidings and being escorted along the ballast from one train to the other by the guard and his red flag.

When the family were spending time in Haworth, I would often travel from there to York or Doncaster, which gave the opportunity to travel on various Class 1 trains between Leeds and York or Doncaster. On the

latter, the HST buffet would open somewhere around Wakefield and I would be one of the first customers to partake in one of my favourite treats. Two slices of "Hot Buttered Toast" was a very affordable 25p and always tasted so much better than anything I ever made at home, especially on a cold day. A chance meeting at Doncaster during one of those trips in April 1987 led to one of a number of sideline interests that I've dipped in and out of over the years. I got chatting to a lad from Sheffield who collected carriage destination labels, which were displayed on carriage droplight windows. There were plenty of named train services at the time, many of which had attractive designs on the labels (or window stickers as we referred to them). We got on well and he shared an innovative idea he'd had, which had been very productive. He would write to the Area Managers at major stations, politely asking for free samples, which would yield parcels of the more desirable unused labels, as opposed to those with sticky ends which had to be carefully peeled off carriage windows. This led to a spell where I wrote around 100 such letters in the months that followed, many of which yielded packages of anything up to 50 labels and other BR paraphernalia. The most exciting part of my weekdays was returning from school to see if my mum had put a large brown envelope on my desk and if so, seeing where it had been franked and exploring its contents. In the same way as football stickers were exchanged between many of my friends, I would carry a large envelope of "swaps" with me when out on the railways, for which there was usually good demand. With having the unused variety and some of the more attractive and obscure examples, sometimes I would sell these and come home a few quid better off!

Around this time, it became clear that the days of the Liverpool to Scarborough and Newcastle TransPennine services being the domain of Class 45s were numbered. Generally I spent more time in and out of Manchester Piccadilly than Victoria, as it saw greater numbers and types of loco-hauled trains, and was more easily reached by the 15-minute train journey from my local station in Sale. In the weeks before the summer timetable began in May 1987, I prioritised these services and my movements on 16 April and 4 May are summarised in Tables 3 & 4 and show what was possible. There's an in-joke today among bashers that 47555 used to turn up everywhere and my first encounter of it was on the second of those dates, when it was interloping on a Class 45 turn, signalling the new order that was to come. After the timetable change that May, the roles were reversed and these TransPennine trains

were hauled by Gateshead based Class 47/4s, with a Peak only making occasional appearances. Class 45 diagrams were then limited to a daily return from Bristol to Weston-super-Mare and a handful of weekend cross-country services. They would sometimes make appearances on routes which they used to dominate, also including the Midland Main Line, but such instances became increasingly uncommon.

Table 3: Moves to make the most of TransPennine Peaks on 16 April 1987

Train	From	To	Mileage
303082	Sale	Manchester Piccadilly	5.75
45149	Manchester Victoria	Stalybridge	7.5
45142	Stalybridge	Manchester Victoria	7.5
45145	Manchester Victoria	Stalybridge	7.5
150226/228/218	Stalybridge	Manchester Victoria	7.5
150236/238/202	Manchester Victoria	Stalybridge	7.5
45136	Stalybridge	Manchester Victoria	7.5
47402	Manchester Victoria	Stalybridge	7.5
142013/014	Stalybridge	Manchester Victoria	7.5
45149	Manchester Victoria	Stalybridge	7.5
DMU	Stalybridge	Manchester Victoria	7.5
EMU	Manchester Piccadilly	Sale	5.75

Table 4: Moves on 4 May 1987

Train	From	To	Mileage
304 016	Sale	Manchester Piccadilly	5.75
45133	Manchester Victoria	Stalybridge	7.5
47555	Stalybridge	Manchester Victoria	7.5
150203/211/219	Manchester Victoria	Bolton	10.75
150245/231/221	Bolton	Manchester Victoria	10.75
142 001	Manchester Victoria	Bolton	10.75
52059/59153/51901	Bolton	Manchester Victoria	10.75
47553	Manchester Piccadilly	Stockport	6
87015	Stockport	Macclesfield	18
86236	Macclesfield	Stockport	18
43126 & 43030	Stockport	Manchester Piccadilly	6
86103	Manchester Piccadilly	Stockport	6
86237	Stockport	Manchester Piccadilly	6

I was back in Blackpool staying with my grandparents during the half term break, when we made an another of our bargain Senior Citizen Railcard trips to London on Wednesday 27 May. Most of the day's traction was relatively straightforward, with 150 235 across the Fylde and 87012 from Preston to Euston. Our return train was the 19.30

Left: Poulton-le-Fylde was one of the stations that would often involve loco-hauled moves to and from when staying with my grandparents in Blackpool. On 15 August 1988, when the freight line to Fleetwood was still in regular use, as seen deviating to the right, 47561 approaches the station in the early stages of its journey to Manchester with the 09.28 departure from Blackpool North.

area from the comfort of a train seat and see what there was to see. "Line bashing" has continued to be a secondary pursuit over the decades that followed and I have made many an exploratory fill-in move to traverse some new track.

I couldn't resist a closer inspection of the many locos on display at Crewe Works, when an open day was held on Saturday 4 July 1987, the first that the Works had held for a number of years. Crewe Works was still very active then, with a huge amount to see, including plenty of Class 37s, mid-way through conversion to the new 37/5 and 37/7 sub-classes and the recently withdrawn and forlorn looking 50011. The first few members of a new class being built there were on display too. BREL had prepared signage to proudly introduce the new Class 87/2, which was being constructed at Crewe. These locos were a development of the existing Class 87 locomotives; however, as they had various features distinguishing them from 87s, including a completely different appearance, when the first examples entered service the following year they were introduced as Class 90s.

With the arrival of July and the warmer weather,

Euston–Blackpool North, which helpfully produced 85013 to Preston and after the engine change there, I walked to the front First Class coach and leaned out of the window to read the number of our diesel traction. As it pulled out of Preston in the darkness nearing midnight, I was pleasantly surprised to take in the sight and sound of Crewe Diesel's "no heat" 47345.

June saw the beginning of another of my sideline railway interests, when during a trip to Liverpool I spontaneously decided to take a Class 508 under the River Mersey and along the New Brighton branch on the Wirral. This was the first instance when I went for a run along a line simply to explore a new

NBs started to turn up a little more frequently. I mustn't have been at school the following week as I was up and down between Manchester, Stockport, Macclesfield and Crewe on the Monday and Tuesday. Haulages included the relatively routine 47444, 47451, 31459, 47629, 86242 and an HST, along with Tinsley's 31119 on the 10.45 Liverpool–Great Yarmouth and the more exotic Old Oak Common based 31163 on the 12.22 Sheffield–Liverpool, both on 6 July.

The following Saturday (11 July) saw me travelling behind six different classes of locomotive between Manchester, Stockport and Macclesfield, making a productive day which involved 81007, 47438, 47475, 43070 & 43041, 86232, 86251, 87002, 87004 and two pairs of 31s on consecutive TransPennine trains. After doing 31469 & 31443 on the 09.45 Liverpool–Sheffield to Stockport and returning to Manchester for the next hourly Class 31 hauled service, I was delighted to find 31161 and 31215 on the 10.45 Liverpool–Yarmouth. The two no-heat locos continued to Norwich, where 31417 then took the train to the East Coast. The following year 31161 changed its identity, when after 31401 was damaged in a collision, its power unit and electrical equipment were transferred to 31161 and it became the 70th member of the class to be fitted with electric train heating equipment. Unusually, instead of being given the number 31470 which would be expected, being the next available number in the 31/4xx series, it was numbered 31400. It returned to traffic in May 1988 and I shared a number of subsequent journeys with the loco.

Above: After my first encounter with the loco in 1987, 31 years would pass before I next had 31163, when it operated at the Chinnor & Princes Risborough Railway's diesel gala on 17 March 2018. At the end of the day, it powered the 16.20 "Thrash-Ex" from Chinnor to Princes Risborough, which at the time was in the process of being connected to the main line station. It is seen there, along with 33207, Class 17 D8568 and 37227 on the rear, during the 2018 "Beast from the East" which was the name given to some terrible Arctic weather conditions, rather than any rare Eastern Region traction!

CHAPTER 3 –
A BUSY SUMMER

The last week of July was the beginning of both the school summer holiday and the time when I started to travel further afield more regularly. Having just completed Year 9 and chosen which subjects I would study from September in the new and as yet untested GCSE qualifications, I could now travel liberated by the knowledge that I would never have another French or History lesson again. On Saturday 25 July 1987, I began the first of the many Rail Rovers that I have travelled on over the years; an East Midlands Rover costing a bargain £13.90, which compares very favourably to the ticket's 2019 child price of £76.90. This gave unlimited rail travel from Sheffield and Stoke-on-Trent across to the East Coast and as far south as Bletchley, Bedford and Peterborough. The only drawback was that the western boundary of validity was Chinley, meaning I had to pay a further £1.20 for a return to the Derbyshire town each day. I travelled nearly 2000 miles that week and this included a decent number of NB haulages, given that I didn't know which trains were loco-hauled, let alone which of these were more likely to provide unusual traction. I also included a couple of bits of line bashing, such as when a Class 115 (51855 & 54491) took me over the Bedford–Bletchley route. Some of the most interesting and varied haulages, however, were had across the Pennines at the start and end of each day; during the week these included 31259 (07.38 Liverpool–Sheffield on 28 July), 47004, 47200, 31412, 31413, 31444, 31454 and 31461. Other locos that transported me that week were 31409, 31413, 31447, 31457, 47618, 47626 and a large number of HST power cars on the Midland Main Line. Three other big winners that week were 31189 & 31296, which I was delighted to stumble across at Nottingham on 26 July (13.10 Norwich–Sheffield) and 45141 which I took south from Sheffield on

Below: On 30 July 1987, 50002 "Superb" was closely followed to Manchester Piccadilly on the final stretch of the 13.54 Oxford–Manchester relief service. Unfortunately the loco was then beyond reach as it returned the coaches to their home in the South East by way of an empty stock working. The chap on the platform may be burying his head in his hands in frustration, perhaps after learning this! *Anthony Flusk*

Above: On 8 August 1987 I found 31405 making a number of trips between Exeter and Barnstaple. A few weeks earlier and shortly before I had it on a Norwich to Sheffield service, 31296 was in action in Devon. The loco is seen at Exeter St. Davids before hauling the delayed 16.45 to Barnstaple on 6 July 1987. *Gray Callaway*

30 July (07.45 Newcastle–Poole). The only blemish that week was later that day, when as I arrived into Platform 4 of Stockport station behind 31454. While my train slowed to a standstill, the highly unusual sight of the rear section of a rake of Network South East livery Mark 1 coaches was leaving the adjacent "down fast" Platform 3. I had just missed whatever was on the front of this, so continued to Piccadilly along the "down slow" line behind 31454. At Manchester, I was amazed to find 50002 on the blocks. It had worked the 13.54 Oxford–Manchester relief, in place of an earlier Poole–Manchester service which had been cancelled. This was the first time I had seen "Superb" and the joy of such a rare find turned to indignation when I learned that it would be returning to its home territory on an empty coaching stock (ECS) working. Had I known that I would end up with a four-figure mileage from 50002, I could have perhaps have been comforted that day.

I was practically insatiable for rail travel that summer and continued chasing locos near and far. Three days after the East Midlands Rover, my Mum took my two brothers and I by train for the day to Aberystwyth and then on the 1 foot 11.75 inch gauge steam railway to Devil's Bridge. I wasn't particularly interested in steam engines, however this was a free day out, courtesy of the family railcard, and the three Vale of Rheidol locos (numbered 7, 8 and 9) were still owned and operated by British Rail then. They were therefore in the Platform 5 Locomotives book and number 7 constituted another haulage to highlight, as it provided us with a scenic journey along some new track.

The Family Railcard was put to good use again a few days later, which meant another expenses paid trip, this time from Manchester to Exeter St. Davids, in order to visit some family friends who lived near Topsham, which is roughly half way along the Exmouth branch. We had been on a number of family holidays to Cornwall before; however, this part of British Rail's Western Region was completely new railway territory for me. Considering that my parents stipulated which trains we travelled on, the outward journey on Saturday 8 August consisted of a very satisfactory majority of diesel haulage, with 47631 from Piccadilly to Crewe (likely to have been the 08.08 Manchester–Weymouth), 86251 to Birmingham New Street and a BR blue liveried pairing of 31422 and 31413 forward for the 165 miles to Exeter. The friends hosting us, collected my parents and brothers from St. Davids station and I was granted permission to further explore the railways. I was intrigued to witness 31405 working the Barnstaple diagram – a Saturdays only turn during the summer of 1987, where a Bristol Bath Road based 31 would make four return trips to North Devon. Being well acquainted with BR Blue liveried 31/4s, a return ticket to Plymouth seemed to be a better way to spend what remained of the afternoon, seeking Western Region Type 4s. With hindsight, however, I wish I'd had the foresight to take a loco-hauled train to Barnstaple when I had the opportunity, especially as it was something I would end up seeking three years later. During my time in Exeter, I remember tuning my ear into the unfamiliar Devon accent, with which the station announcer would broadcast "Exeter Saint Davids, this is Exeter Saint Davids" to the new arrivals on each train. Cardiff Canton

based 47565 took me west and I was bowled over by the scenery, with hoards of holidaymakers on the beaches and sea wall which hugged the line at Dawlish and rolling hills of Western Devon. I didn't have the luxury of being able to wait and chose the traction back to Exeter and Crewe Diesel based 47463 did the honours, which whilst an everyday engine for the visitor from Manchester, was at least a new haulage. I made my way to Topsham for the evening meal, courtesy of a first generation DMU which I didn't bother noting the number of, as had started to become my habit. We only stayed the one night in Devon and over breakfast the next morning, our perceptive hosts asked whether I would like to return and stay again to see more of the region's trains. I didn't need any further persuasion and as 47626 transferred our family directly from Exeter to Crewe (followed by 86212 to Manchester), I was already looking forward to returning.

Then followed one day which didn't involve any trains and most likely consisted of updating my records with the recent new loco haulages

Table 5: Moves on 11 August 1987

Traction	From	To	Mileage	Service
Unit	Sale	Manchester Piccadilly	5.75	
47488	Manchester Piccadilly	Stockport	6.00	08.08 Manchester–Poole
303082	Stockport	Manchester Piccadilly	6.00	
31432	Manchester Piccadilly	Stockport	6.00	
47277	Stockport	Manchester Victoria	11.75	07.18 Nottingham–Blackpool
47402	Manchester Victoria	Stalybridge	7.50	xxxx Liverpool–Newcastle
150 256+150 220	Stalybridge	Manchester Victoria	7.50	
86225	Manchester Piccadilly	Macclesfield	18.00	
87011	Macclesfield	Manchester Piccadilly	18.00	
86237	Manchester Piccadilly	Stockport	6.00	
47441	Stockport	Manchester Piccadilly	6.00	06.25 Poole–Manchester
47588	Manchester Piccadilly	Stockport	6.00	
31419	Stockport	Manchester Piccadilly	6.00	
86101	Manchester Piccadilly	Stockport	6.00	
31421	Stockport	Manchester Piccadilly	6.00	
86221	Manchester Piccadilly	Stockport	6.00	
43185 & 43124	Stockport	Manchester Piccadilly	6.00	
Unit	Manchester Piccadilly	Stockport	6.00	
47609	Stockport	Manchester Piccadilly	6.00	07.20 Harwich–Manchester
Unit	Manchester Piccadilly	Sale	5.75	

Table 6: Moves on 27 Aug 1987

Traction	From	To	Mileage	Service
EMU	Sale	Manchester Piccadilly	5.75	
47418	Manchester Victoria	Stalybridge	7.50	xxxx Liverpool–Newcastle
150238/206	Stalybridge	Manchester Victoria	7.50	
150222	Manchester Victoria	Stalybridge	7.50	xxxx Liverpool–Newcastle
47401	Stalybridge	Manchester Victoria	7.50	
47417	Manchester Victoria	Stalybridge	7.50	xxxx Newcastle–Liverpool
DMU	Stalybridge	Stockport	7.50	
86438	Stockport	Manchester Piccadilly	6.00	
47631	Manchester Piccadilly	Stockport	6.00	
86426	Stockport	Manchester Piccadilly	6.00	
31460	Manchester Piccadilly	Stockport	6.00	xxxx Liverpool–Sheffield
31407	Stockport	Manchester Piccadilly	6.00	xxxx Sheffield–Liverpool
85003	Manchester Piccadilly	Stockport	6.00	
86419	Stockport	Manchester Piccadilly	6.00	
31432	Manchester Piccadilly	Stockport	6.00	xxxx Liverpool–Sheffield
43010 & 43025	Stockport	Manchester Piccadilly	6.00	
31452	Manchester Piccadilly	Oxford Road	0.50	xxxx Sheffield–Liverpool
EMU	Oxford Road	Sale	5.75	

and sightings, seeing friends or riding my bike somewhere. I was out again the next morning for some routine but productive moves to familiar haunts in Greater Manchester. Tables 5 and 6 show the variety of locos that could typically be found passing through Manchester, with details of the services being worked, where they are known. It was common for services to be worked by different engines on consecutive days, meaning there was a practically unending variety of haulage on offer. I would take the odd diesel or electric multiple unit move to position myself for the next inward or outward loco-hauled train, but on the whole that wasn't necessary as there

were so many "proper trains" to choose from. My new haulage success rate was somewhere in the region of 50% at this time, so there was plenty to be going on with, without my limited funds having to bear the higher costs of travelling further.

The next long-distance trip was a family holiday to Tighnabruaich on the Cowal Peninsula overlooking the Kyles of Bute in Western Scotland during August. The journey there was interesting, not only because it involved taking our car across the Firth of Clyde on the ferry from Gourock to Dunoon, but due to the deviation I had persuaded my parents to make. Knowing there was a large diesel depot at Eastfield,

Table 7: Locomotives present at Eastfield Depot on Saturday 15 August 1987:

Shunters	08852
Class 20	20114 & 20138
Class 26	26006, 26008, 26015 & 26046
Class 27	27003, 27005, 27007, 27008, 27024, 27046, 27050, 27054, 27056, 27066 & 27207 *
Class 37	37004, 37014, 37084, 37191, 37406, 37407, 37408, 37410, 37411 & 37424
Class 47	47004, 47109, 47118, 47137, 47209, 47595, 47633, 47661, 47701, 47709 & 47712

* 27207 had the incorrectly applied departmental number ADB68025, which should instead have read ADB968025.

to the north of Glasgow which would be full of unfamiliar Scottish traction, I had asked if we could "have a look" there. My sympathetic parents and brothers, presumably knowing that I had becoming increasingly mad about diesel engines, kindly agreed to make the necessary detour. I vividly remember walking inside the shed building, where my Dad found the supervisor, who happily gave us free reign of the depot. It was eerily quiet as we walked up and down the roads of stabled locos outside and inside the shed building. This wasn't a loco bashing day; however, getting up close and personal with Scottish Region Class 20s, 26s, 27s and split-headcode Class 37s at track level is inspiring enough to warrant a mention. For posterity, the engines present at Eastfield depot on Saturday 15 August 1987 are shown in Table 7. I would have been more than happy

to take a train hauled by any one of these; however, I did in future manage to do so with 27056, the six Class 37/4s, 47004, 47209, 47661, 47709 and 47712.

After a pleasant week of exploring the peaceful environs of the Kyles and Isle of Bute, we returned by car from Scotland on Saturday 22 August and the next morning I was on a train heading back to Devon. I took 86426 to Stockport, 47474 forward to Birmingham and HST power cars 43178 & 43067 on to Bristol via Gloucester, where the recently withdrawn 50006 had been unceremoniously deposited. I paused for a while at Temple Meads, as there was plenty going on at the station and the adjacent Bath Road depot, which between them saw multiple examples of classes 08, 31, 33, 37, 43, 45, 47 and 50. Bristol's visitors that afternoon included the unique 47901, which had been renumbered from 47601 in 1979 when it was fitted

Below: One of the large logo liveried Scottish locos viewed at Eastfield depot on 15 August 1987 was 37424. At the time no one knew what a colourful future it would have; after being withdrawn in 2000, its condition declined over subsequent years and the Type 3 entered preservation in 2008 in a poor state of repair. It was acquired by DRS some years later and benefitted from a complete rebuild, returning to main line service with the fictional number 37558 in 2016. It is shown at Altrincham with 37407 when working The Branch Line Society's "Nosey Peaker" railtour on 14 June 2018.

Above: The "Freedom of the North-East Rover" allowed unlimited travel on the Settle to Carlisle Line, which was still threatened with closure in 1988. This triggered larger numbers of enthusiasts and the public alike to travel on it. On 12 November 1988, 47413 attracts some interest at Appleby, while working the 12.42 Carlisle–Leeds. *Richard Allen*

with a Ruston Paxman RP12RK3CT engine. This was to create the test bed for the Class 58s and the last of this new class had since entered service a few months earlier in March 1987. Prior to that, the Class 47 had been selected as a test bed for the planned Class 56s after derailment damage in 1974 necessitated repairs. When these were carried out, it had its standard Sulzer 12LDA28C engine replaced with a Ruston 16RK3CT. It was renumbered to 47601 in 1975, as the ETH converted locos had only been renumbered as far as 47555, giving the experimental loco a unique number range at the time. This had to be extended to 47901 when it received its second new engine, as further ETH converted locos had then reached the 476xx range. It generally wasn't considered a "real 47" by Class 47 bashers once it carried a non-standard power unit, as the engine is of course the heart of a loco, producing the all-important sound which characterises a class. I also noted 31294 working an unknown passenger service that afternoon in Bristol, a loco that I didn't have another chance to travel behind before it was withdrawn in 1995. Its classmate, 31427, however conveyed me from Bristol to Exeter, where I was then met by my hosts.

The road between Exeter and Topsham runs along the northern bank on the lower reaches of the River Exe, before it widens into the estuary which meets the sea at Exmouth. The main line from Exeter to Dawlish and beyond also follows the southern banks of the Exe, with the three modes of transport, which represent three different historic eras, all running parallel for a few miles. As I was being driven from Topsham to Exeter the next morning, I vividly remember the fine sight of a large logo Class 50 with a corresponding blue liveried rake of coaches comfortably outrunning us on the opposite side of the river about half a mile away, as if it was the warm up act for the day's performances. These then consisted of 50004 from Exeter to Plymouth, where I wandered off to Laira depot to see how many 50s I could find tucked away there, before my first run behind 50007 back to Exeter.

I made the best possible use of my return ticket from Exeter to Manchester on Wednesday 26 August 1987, simply waiting for what I thought was interesting enough traction, then taking that forward to the next station. The first winning loco of the day was 50009 from Exeter to Taunton on the 09.55

Above: Unlike the mystery Class 31 that took me down the coast to Sunderland and Middlesbrough on 3 September 1987, the identity of the Type 2 retracing its steps on 26 May 1989 is known. 31444 leaves Newcastle with the 13.30 to Middlesbrough, which on 26 May 1989 was formed of three coaches. Richard Allen.

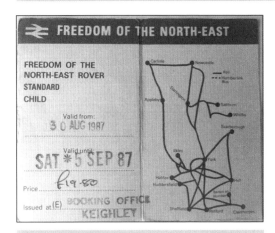

Above: The 1987 Freedom of the North-East Rover, complete with a map showing its validity, before tickets were standardised in the late 1980s.

Paignton–Paddington, then I picked up red stripe railfreight liveried 31301 to Bristol on the 11.33 Exeter–Leeds relief service. I watched the comings and goings at Bristol Temple Meads until another Western Region loco arrived on a Birmingham-bound train. That turned out to be 50008 "Thunderer" on the 10.20 Plymouth–Aberdeen, which returned me to the Midlands via Gloucester.

Two significant conversations took place on that fine August day, which helpfully steered my interest. A secondary advantages of always sitting in the front coach to listen to the locomotive's thrash, is that you can overhear or join in with the conversations between fellow enthusiasts, picking up information, stories and the jargon of the hobby. It was as 31301 rumbled through Somerset on its relief working that I first heard the word bashing being used, as I listened in on a discussion about individuals' preferred loco classes. That was the point when I realised that is what I had become – a basher, one that enjoys journeying behind diesel locos, accruing haulages and savouring the sound of their engines at work. An hour or so later, I was chatting to one of Bristol's enthusiasts on one of Temple Meads' platforms and he made a very good case as to why I should buy a camera. He persuaded me that it's important to not only make written records of what I saw and did, but pictorial records too. It took until Christmas, four months later, before I was in possession of a camera and he was absolutely right. I owe him a drink for labouring that point to me and my only regret is not spending more on better quality film and upgrading the camera sooner. In all honesty though, when all aspects of my hobby had to be funded from the proceeds of my paper round, a better quality camera and film would have been at the expense of travelling as much, which simply wasn't an option, as haulages and diesel mileage were my priority.

After a brief two-day break without any rail travel, I was back in the small family holiday home in Haworth for a week or so. Instead of joining my parents and brothers exploring the sights, architecture and attractions of Yorkshire, my sympathetic parents sanctioned another pass out for their teenage traveller. The nearest main line station, Keighley, wasn't far away, so I was well positioned to make the most of a "Freedom of the North-East Rover" which gave unlimited rail travel as far as Carlisle, Newcastle, Cleethorpes and Sheffield. I didn't make any advance plans and simply spent the week travelling the region, much of which was new to me and seeing and doing whatever turned up. Some of the highlights of that week included taking 47488 on the length of the Settle and Carlisle route, scoring 45115 on the 14.40 Newcastle–Poole on 30 August and 47270 on the 15.47 York–Bristol relief service on 31 August. This relief working ran a number of days that week and was non-stop between York and Sheffield, taking the route via Pontefract

Baghill, which was uncommon for loco-hauled trains. On the Sunday, instead of using the usual route via Darlington, all East Coast Main Line services were diverted through Eaglescliffe, at which some trains were making scheduled stops. I therefore made an unusual HST move (43031 & 43135) from Leeds to Eaglescliffe and took the opportunity to walk to the now closed Thornaby depot and see which locos were present. I was pleasantly surprised to find 11 out of the 19 newly converted Class 37/5s (only 37501–37519 had been completed at that point), with 37501–502, 505–507, 509 & 515–519 all present. Many of the recently introduced Pacer units had gearbox problems and had been removed from service that summer, leading to loco-hauled trains temporarily being used on various routes across the north of England. During the afternoon of Thursday 3 September 1987 while at Newcastle, I was so pleased to stumble upon a blue liveried Class 31 on one such train, heading for Middlesbrough via Sunderland, that I didn't make a note of the loco's number; it is likely to have been a 31/1, quite possibly one which I didn't have again and my efforts to establish which loco this was, have so far been unsuccessful. If any readers can confirm the identity of this 31, I would be delighted to hear from them!

The end of that North-East rover marked the end of a six week school summer holiday, consisting of little other than travelling by rail. Excluding the eight days which the family holiday to Scotland occupied, I was chasing locos on all but eight of the possible 44 days and had notched up a whopping 154 different haulages, as Table 8 summarises.

Table 8: The great haulage haul of the school holidays, 25 July – 5 September 1987

Class	Loco Haulages
31	31189, 31259, 31296, 31301, 31407, 31409, 31410, 31412, 31413, 31419, 31421, 31422, 31425, 31427, 31432, 31435, 31442, 31444, 31447, 31452, 31454, 31457, 31460 & 31461
HST	43008, 43010, 43011, 43013, 43017, 43020, 43025, 43027, 43031, 43032, 43035, 43038, 43039, 43040, 43043, 43044, 43046, 43049, 43050, 43051, 43052, 43056, 43059, 43060, 43061, 43067, 43069, 43072, 43073, 43078, 43079, 43080, 43081, 43082, 43083, 43088, 43089, 43095, 43104, 43105, 43120, 43124, 43135, 43138, 43141, 43145, 43152, 43153, 43155, 43156, 43157, 43158, 43160, 43162, 43164, 43166, 43178, 43179, 43183, 43185, 43188, 43190, 43195 & 43198
45	45115 & 45141
47	47004, 47147, 47200, 47270, 47277, 47377, 47401, 47402, 47417, 47418, 47441, 47463, 47467, 47474, 47488, 47489, 47491, 47497, 47536, 47557, 47565, 47588, 47600, 47608, 47609, 47618, 47623, 47626, 47631 & 47659
50	50004, 50007, 50008 & 50009
85-87	85003, 85028, 86101, 86212, 86213, 86221, 86225, 86231, 86237, 86238, 86251, 86253, 86257, 86411, 86412, 86419, 86426, 86427, 86430, 86438, 86439, 87003, 87009, 87011, 87013, 87021, 87027, 87029 & 87032
Steam	7

CHAPTER 4 –
INCREASING TRACTION

As summer morphed into autumn, the routine of school and homework on weekdays returned, generally restricting my outings to Saturdays during term-time. At that time British Rail operated two timetables; what enthusiasts referred to as the Summer Timetable ran from mid-May until late September or early-October, with the Winter Timetable covering the alternate period. Locos without train heating equipment could be found on a number of the summer workings, much to the satisfaction of bashers, but from the time the British weather began to decline in September, these became less common. I happened to be in the right place at the right time for two such workings on 12 September 1987, when I added 31158 (07.38 Liverpool–Sheffield) and 31142 (09.45 Liverpool–Sheffield) to my haulage tally. The October half term gave me the opportunity to stretch my legs more, when I spent

seven days travelling on a Coast and Peaks Rover, which provided unlimited travel in the area bordered by Sheffield, Manchester, Liverpool, North Wales and as far south as Tamworth and Stoke-on-Trent. Day one was Saturday 24 October, which started well with five consecutive winning locos; 47559 from Manchester to Crewe (08.08 to Poole), 87019 to Tamworth and 86252 back to Crewe, where I found the unusual instance of a pair of 47s at the front of a Euston–Holyhead train. As both 47512 and 47538 were required, I took these to Chester, where I then decided to do a bit of line bashing, taking 142 011 on the then non-electrified route to Hooton and 508 120 through to West Kirby and back to Liverpool.

After making my way to the upper level of Lime Street station, I was pleasantly surprised to find 45128 on the front of the 15.03 to Newcastle. Class 45s on the Liverpool to Newcastle route had been ousted by

Below: Repeating the move I made on 24 October 1987, when 45 haulage on the TransPennine route had become rare, on 18 June 1988 45128 again prepares to depart from Liverpool Lime Street, this time with the 16.58 to Newcastle. The Peak was withdrawn less than a year after this photograph was taken and was subsequently scrapped. *Anthony Flusk*

Above: Nine months after my very first Class 37 haulage from this station, 37431 accepts the road as it leaves Church Stretton on the 09.15 Liverpool Lime Street–Cardiff Central on 25 July 1988.

Left: This ticket must have represented the best possible haulage to the pound ratio in 1987. For 15p, the outward leg of this ticket provided 17 miles behind 47436 from Keighley to Leeds and a further 17.25 miles behind 47417 from Leeds to Huddersfield.

47s the previous May; however, they continued to make occasional guest appearances on the route. Knowing that this was now quite a rare working, I walked to the front of the 7-coach rake of Mark 2 open coaches and found a seat close to the Peak. Before the departure, after inspecting the train's traction, a chap came up to the carriage window and shouted "Wagon basher!" at me. I had been around long enough to know most of the nicknames given to the various locos, with most classes having a derogatory term, Wagon being the one for Peaks. Bashers are often guilty of being very territorial in what they like, don't like, do and won't do and whilst I have partaken in some entry level behaviour of this type, diesel haulage has always been my priority, irrespective of whether it's a class I am more or less fond of. I took 45128 from Liverpool to Stalybridge and unlike the previous time I had it six months earlier, it refrained from setting itself on fire and kept to time! This turned out to be the final time I had a Class 45 on any section of the TransPennine route and there weren't many subsequent dates when they could be found on this once common stomping ground.

A couple of days later, I ran into the lad I had met in Doncaster earlier in the year, who had tipped me off about writing to stations to request carriage window labels. It turned out he and his friend were also spending their half term break on a Coast and

Peaks Rover and the three of us spent much of the rest of the week together. A few days later, while sitting on a Euston-bound train at Liverpool Lime Street, we saw Justine Kerrigan and Sean McKee who played screen couple Tracy Corkhill and Jamie Henderson in Brookside, walk past and enter the First Class coaches. The Liverpool-based soap opera was massive in the 1980s, attracting nearly ten million viewers, which included the three of us. We spent the journey to Crewe deliberating whether it would be worthwhile or embarrassing to go and speak to them. In the end, curiosity won and two autographs provided some interesting variety to the list of locomotive moves in our notebooks!

The rest of the week was spent knocking about the North West and North Wales, mainly chasing diesel locos, with the North Wales Coast route between Crewe and Holyhead providing a rich vein of Class 47 haulage. There were regular services on this route between Euston and Holyhead, which connected with ferries to and from Ireland, plus a smaller number of trains which continued to Llandudno or Cardiff. The AC electric locos would hand over to diesels at Crewe and diesel-hauled services on the route west of Crewe required at least seven different 47s to work each day. I finished those seven days of travel with a small but significant footnote to my bashing chronicles. After taking 47619 from Crewe to Church Stretton on a Crewe–Cardiff service, I was positioned for the corresponding northbound working, which I hoped would be one of the six Cardiff-based 37/4s (37426–37431) that had been fitted with ETH equipment and renumbered the previous year. Surprisingly, I hadn't had any 37s at that point and was pleased to open my English Electric Type 3 account with 37430 back to Crewe on 30 October 1987. That was the first of many "Tractor" moves, which over time took me across the length of the country, firstly on service trains and in latter years on railtours and heritage railways.

I reverted to the pattern of bashing on Saturdays and the occasional Sunday, mostly around the North West, through November and until school finished for Christmas. I then made my way up to Blackpool to stay with my grandparents for a few days immediately before Christmas and had a couple of trips out, with mixed success. During a day trip from Blackpool to Carlisle, courtesy of another £1 child ticket with the Senior Citizen Railcard holders, while my grandparents explored the city's architecture and culture, I remained at Citadel station. Until mid-afternoon it was the usual combination of electric locos on the West Coast Main Line, 47/4s on the Leeds

and Glasgow via Kilmarnock trains and DMUs on services to Barrow and Newcastle. Shortly before we were due to return south, no heat (in late December!) 37098 rolled in on a DMU-replacement service from Newcastle. I knew it was rare and was desperate to do this on the return Eastbound working, but that would have meant getting a later train back to Preston. I could persuade my Grandma to do a lot of things, but she was having none of it that day. I can still hear her voice telling me sternly "I don't care how rare it is, we're not taking that train anywhere", so I had to settle for watching it depart without me and remember a number of satisfied 37 fans poking their heads out of the sliding windows within the Mark 1 stock as it departed. 37098 lasted another decade in service and worked a handful of passenger trains during that time, but none that I managed to cross paths with. As I have learned with bashing, you win some and you lose some.

I returned from Blackpool to Manchester by train on Christmas Eve and instead of taking one of the direct trains formed of a comparatively bland DMU, I carved out a more creative loco-hauled journey. Starting with 47647 (on either 09.50 Blackpool–Euston or 11.21 Blackpool–Nottingham), I then took 45107 from Preston to Manchester Victoria. I have no idea of which working the Peak was employed on that 24th December; it was around late morning and I have established that shortly after this, at around 13.30, it brought an empty set of coaches from Longsight to Manchester Piccadilly.

There were no services on Christmas Day or Boxing Day at that time, not that I would have dared to ask to leave the family home on those special days, or even wanted to forfeit the pleasures of exchanging presents and my Mum's home-made Christmas dinner and Christmas cake. Went spent a few days in Haworth between Christmas and New Year and I exchanged being squeezed in the back seat of our average sized family car with my two brothers, for the longer train journey from Sale to Haworth. This gave me the opportunity to fit in 47538 on the Leeds to Keighley section after taking 47553 on a deviation to Wakefield. One week after 45107 was my previous winning Peak, I was delighted to come across another on the final day of the year, which provided me with a first and a last. As 45113 arrived at Leeds on the 07.45 Newcastle–Poole, I was now equipped with the highly useful Christmas present of a new camera, making this the first loco-haulage that I could document on film. I knew their days were numbered, but I didn't know at the time that this would turn out to be the

Above: At first glance the train arriving into Newcastle appears to be one of the regular services from Liverpool, as it's comprised of the TransPennine liveried coaches used on the route at the time. 47424, which carried the name "The Brontes of Haworth" between 1988 and 1991, is, however, standing in for a DMU on the 17.00 from Middlesbrough on 20 May 1989. *Richard Allen*

last time I scored a Peak on a service train. That brought me to the end of the year, during which I was hauled by 347 different locos, the majority of which were new to me and these consisted of 10 x Class 31/1s, 45 x Class 31/4s, 1 x Class 37, 91 different HST power cars, 14 x Class 45s, 93 x Class 47s, 5 x Class 50s, 1 x Class 81s, 7 x Class 85s, 53 x Class 86s and 27 x Class 87s. We stayed in Yorkshire over New Year and as the sun set on 1987, I could look back at one prolific year of loco haulage and look forward to another, which would involve greater numbers, new classes and new destinations.

CHAPTER 5 –
MORE VARIETY

By the start of 1988, the proliferation of second generation DMUs was well under way. The last batch of Pacers (Classes 140–144) had recently been constructed and had replaced their first generation counterparts on the Cornish branch lines and the suburban routes of Greater Manchester, West Yorkshire and the North-East. The first of the Sprinters (Class 150) had entered service on medium distance routes, such as Derby to Crewe and various services across the North, Midlands and Wales. In addition, hundreds of further "Super Sprinters" were already under construction (Classes 155 and 156) or in the planning and design stage (Classes 158 and 159). It started to emerge that many of these would be replacing the reason for the basher's existence – loco-hauled trains. In response, I formed the somewhat misguided SSS (Scrap a Sprinter Society), which a number of my school friends joined and at its height, membership was somewhere in the region of eight people. To me, second generation multiple units

have always been an unwelcome reminder that they replaced heritage traction; however, when I'm being objective, I can see that the efficiencies they brought may have prevented some lines from closing during the railways' dark days of the 1980s and early 1990s.

There was a sense of "do them while you can", not only for the trains that had become a staple ingredient of my haulage diet such as Class 31s on the Liverpool to Sheffield and Cleethorpes trains or Class 47s on the Leeds–Carlisle route, but also for those that I had yet to explore. The latter included the Class 33 hauled strongholds of Bristol to Portsmouth and Waterloo to Yeovil, Eastfield and Inverness based 37s on the West Highland, Kyle of Lochalsh and Far North Lines, 47/7s on the Glasgow–Edinburgh push-pull trains and various routes where Class 50s still worked. In addition, I had already witnessed the final withdrawals and eradication of the Class 25s and 40s and it was clear that 45s weren't far behind. All of this suggested that times were changing and some of

Above: A number of Class 50s worked to Manchester during the 1987/88 winter timetable and Saturday 30 January 1988 was the only date that I enjoyed this rarity. After arriving on the 14.01 from Manchester, 50007 pauses at Stafford on the train in question, before taking charge of the 16.09 to Paddington. *Anthony Flusk*

the everyday sights and sounds wouldn't last for long. There were still hundreds of loco-hauled trains and plenty of classes to choose from every day and even though I now did two paper rounds, the proceeds of these would only take me so far.

I started 1988 with some regular Saturday home fixtures around Greater Manchester and the new haulage success rate was starting to drop. At the turn of the year I'd had 54 of the 69 Class 31/4s and I added four to this in January and another four in February. Excluding 31436, which was withdrawn in 1986, this left five of the sub-class, meaning the level of difficulty and satisfaction involved in tracking down a winning example were increasing in equal measure.

There was a Monday–Saturday Paddington–Manchester diagram through the winter 1987/88 timetable period. The loco worked for 217 miles on the outward 07.03 from London via Birmingham, returning by way of the 14.01 Piccadilly–Stafford and the 16.09 Stafford–Paddington. This was booked for a 47/4, but unbeknown to me, every now and then it would produce a Class 50. There was therefore much excitement among the northern neds when the completely unexpected 50007 arrived into Stockport on the "down" working on Saturday 30 January 1988.

Most of the locals had never had a 50 before, so I had mixed feelings about this, it being one of only five 50s that I didn't need! The rarity value and my appreciation of the class more than made up for that, though. With the loco being one of my first 50s, this rare working into my hometown and it lasting until the class's final working for BR, "The 50 Terminator" which I travelled on six years later in 1994, 50007 has always been one of my favourites. It was initially named "Hercules" in 1978 and in early 1984 it was renamed "SIR EDWARD ELGAR" and repainted into Brunswick Green to mark the 150th anniversary of the Great Western Railway. It carried this name and livery for 30 years, until it was renamed back to "Hercules" in 2014. Many 50 bashers disliked the name and livery, which deviated from the blue colours and warship name theme carried by the rest of the class, but for me it was the only version I knew for 27 years and I don't mind a bit of variety.

Known instances of Class 50s working to Manchester during the winter 1987/88 timetable are shown in Table 9. Apart from 50007 on 30 January and on 2 April when I was otherwise occupied with a railtour, these were all weekday workings, for most of which I was obliged to be at school.

Below: There were just three daily trains in each direction between Leeds and Carlisle during the late 1980s, formed of ample lengths of Mark I coaches. On 1 June 1988, 47648 pauses at Keighley with the 10.45 Leeds–Carlisle.

Table 9: Dates of known Class 50 workings on the 07.03 Paddington–Manchester, 14.01 Manchester–Stafford and 16.09 Stafford–Paddington during the 1987/88 winter timetable.

Loco	Date loco worked
50020	Tuesday 13 October 1987
50007	Saturday 30 January 1988
50044	Tuesday 9 February 1988
50024	Wednesday 10 February 1988
50041	Tuesday 16 February 1988
50044	Wednesday 9 March 1988
50046	Tuesday 29 March 1988
50007	Saturday 2 April 1988
50027	Thursday 21 April 1988
50036	Monday 9 May 1988

The week-long half term break was spent on another East Midlands Rover, from 15 to 21 February inclusive. This provided the opportunity to start and finish each day behind a Class 31 between Manchester and Sheffield, or continue as far as Peterborough on the daily 10.45 Liverpool–Yarmouth. There were also 47s on the Newcastle–Poole and Bristol–York routes, plus plenty of HSTs which continue to ply their trade on the Midland Main Line today. For me, the most interesting working of that week was the Monday–Saturday Class 45 hauled service between Leeds and Nottingham. It was a fairly solid Class 45 turn, although a 31 or 47 would sometimes be found on it. The diagram is shown in Table 10 and provided me with the opportunity to travel behind 45128, 45115 and 45110 on 15, 18 and 19 February respectively. A fourth loco, 45106, also worked these trains on 17 February, but I didn't manage to catch up with it. None of the four Peaks were required locos; however, that didn't seem to matter in the context of the "How much longer will these locos last?" factor. I remember being surprised how quiet Nottingham station was while the Peaks ran round the life-expired Mark 1 stock that they brought into Platform 1. Few people seemed to be on these trains during the daytime and I expected to see more bashers, given

the type's finite life expectancy. I also noted 45104 at Derby on Saturday 20 February, when it worked a Footex, transporting Derby County fans south to watch their team play against former First Division rivals Oxford United.

Table 10: Monday to Saturday Class 45/1 Diagram during 1987/88 Timetable Period

Headcode	Service
1M03	06.18 Sheffield–Nottingham
1E31	07.25 Nottingham–Leeds
1M45	10.27 Leeds–Nottingham
1E41	13.25 Nottingham–Leeds
1M66	16.27 Leeds–Nottingham
2F25	18.54 Nottingham–Leicester
1E26	22.10 Leicester–Sheffield

By March it was known that the Class 31 hauled trains between Liverpool and Sheffield, which I felt I had grown up with, would be replaced by the new Class 156 Sprinters when the summer timetable period began in May. As this was only a matter of weeks away, I, along with many of the regular North-West based bashers, made a concerted effort to make the most of these and track down some of the last five ETH-fitted examples that I needed on Saturdays 12, 19 and 26 March. Table 11 shows my efforts on the first of these dates. Across those three Saturdays, 16 different Class 31s worked through Manchester and I also managed haulages from 43009, 43031, 43125, 43142, 43181, 43188, 47426, 47434, 47466, 47519, 47582, 47655, 86102, 86206, 86401, 86402, 86411, 87008, 87020 & 87032. The success rate at finding new 31/4s was close to zero, as the only debatable new haulage was 31418, which I boarded at Stockport on 12 March. As mentioned in the Introduction, this moved forward five metres before it failed and the train was terminated. That was academic, however, as I finally caught up with 31418 later that month on another Sheffield–Liverpool service and to make absolutely sure I did it out of Liverpool on the 16.00 to Norwich the following summer and from Manchester Victoria to Daisy Hill in 1992.

Table 11: Moves on 12 March 1988

Traction	From	To	Scored	Working
31463	Oxford Road	Stockport		07.38 Liverpool–Sheffield
31438	Stockport	Manchester Piccadilly		08.22 Sheffield–Liverpool
31448	Manchester Piccadilly	Stockport		08.45 Liverpool–Sheffield
31410	Stockport	Oxford Road		09.22 Sheffield–Liverpool
31454	Oxford Road	Stockport		09.45 Liverpool–Sheffield
31463	Stockport	Oxford Road		10.22 Sheffield–Liverpool
31434	Oxford Road	Stockport		10.45 Liverpool–Sheffield
86102	Stockport	Wilmslow	Yes	12.00 Manchester–Euston
47426	Wilmslow	Stockport		xxxx Cardiff–Manchester
47434	Stockport	Manchester Piccadilly	Yes	xxxx Poole–Manchester
31454	Manchester Piccadilly	Oxford Road		12.22 Sheffield–Liverpool
31410	Oxford Road	Manchester Piccadilly		12.45 Liverpool–Sheffield
47519	Manchester Piccadilly	Stockport	Yes	14.01 Manchester–Stafford
31424	Stockport	Oxford Road		13.22 Sheffield–Liverpool
31463	Oxford Road	Manchester Piccadilly		13.45 Liverpool–Sheffield
Unit	Manchester Piccadilly	Stockport		Unknown
31418	Stockport	Stockport	Yes	14.22 Sheffield–Liverpool
31410	Stockport	Oxford Road		15.22 Sheffield–Liverpool

While I was at Stockport on 12 March, I witnessed the strange and colourful combination of 40122 in grey undercoat, with InterCity livery 97252 (formerly 25314) tucked behind supplying ETH, on a rake of Network SouthEast branded coaches. The 40 had recently worked its final service train, when it powered the 09.30 Euston–Holyhead and 16.15 return working between Crewe and Holyhead on 9 February. It was mid-way through a series of farewell tours marking the end of the class's 30-year career with British Rail and apparently there hadn't been time to finish the repaint for this tour, which ran from Euston to Crewe via Hebden Bridge and Huddersfield. I had decided to play my part in the festivities by booking onto my first railtour, Hertfordshire Railtours' "Tubular Belle", which ran on 2 April. 40122 returned to the class's former stomping ground from Crewe, where I picked it up, along the North Wales Coast route to Llandudno and on to Holyhead, plus a scenic trip to Blaenau Ffestiniog. Wanting to make the most of my last stretch behind the Whistler and the rarity of traversing Chester's centre road that day, I had a bellow as we passed through the station. I was promptly given a stern telling off by one of the stewards, undoing any suggestion that clamping down on such behaviour is a product of recent, increasingly safety conscious times. The following Saturday saw 40122 on its penultimate duty, working "The Vulcaneer" from Manchester Victoria to London Liverpool Street. After an ECS test run between Stratford and Cambridge on Monday 11 April, the final strike of the axe was swung on Saturday 16 April, when the loco retraced its inaugural run from London Liverpool Street to Norwich 30 years earlier and then continued to York in preparation for retirement inside the National Railway Museum, where it can still be found today.

At the end of an enjoyable day in North Wales with 40122, I made my way up to Blackpool and exchanged the company of diesels for that of my relatives, spending Easter Sunday and Monday with my grandparents. The Tuesday consisted of a short hop from Blackpool to Preston and back. Naturally I selected diesel locos over the majority of DMU-operated trains across The Fylde, taking 47434 outward on the 11.20 Blackpool–Nottingham and 47509 back on the 12.00 Euston–Blackpool. In between I made a couple of trips up to Lancaster,

Above: My very first railtour, but one of the last of the multitude of workings that Class 40s carried out for British Rail. On 2 April 1988 and in very damp conditions, 40122 takes a break during a photostop at Betws-y-Coed on the Blaenau Ffestiniog branch, while working "The Tubular Belle". Upon reaching Llandudno Junction, it would take a left turn, continuing to Holyhead and later returning to Crewe. *Anthony Flusk*

adding two new 86 haulages to the book. I rarely sought electric loco haulages and simply noted them as they took place, with the exception of 81s and 85s as they were less common and were affectionately known as "Roarers", due to the louder sound they emitted. Scoring 86224 and 86260 that afternoon took my total for the class to 75 of the 100 built, making it a productive afternoon, as when one reaches the three quarter mark for a class, winning locos usually only come one at a time.

The Senior Citizen Railcard was put to good use the following day, when the three inter-generational travellers made a day trip from Blackpool to Edinburgh. After taking us from Preston to Carstairs on the 07.25 Birmingham–Glasgow & Edinburgh, 86419 continued to Glasgow on the front half of the train, while 47463 was attached to the rear and took us in the opposite direction to the Scottish Capital. While my Gran and Grandad soaked up the Royal Mile and beyond, I made the most of a £2.15 child return to Glasgow Queen Street and a 50p return to Haymarket. This yielded 47703, 47707, 47708,

47711, 47715 and 47716 on the Glasgow–Edinburgh pull-pull services and 47541 and 47482 on the two Euston–Inverness "Clansman" services, which passed each other within an hour at Edinburgh. 47471 was my eighth member of the class that day, when it returned us to Carstairs on what turned out to be the last time I had a diesel on the "Carstairs portion" before the route was electrified in 1991.

I was back to home turf the next morning and proved the theory that if you do enough loco-hauled trains, sooner or later something rare will turn up. I started the day early with 31427 from Blackpool on the 07.20 to Harwich, instead of the Class 47 that this train usually saw. This took me to Stockport via Manchester Victoria and sandwiched between various Class 31, 47, 86 and 87 haulages, I was delighted to find the required 31428 on the 09.45 Liverpool–Sheffield. The bad news was that 31401, which was one of the few other 31/4s I needed, had been withdrawn just a few days earlier, making it the second member of the sub-class to be taken out of service before I'd had it (31436 was other one and was withdrawn in 1986). 31401 was one of the lower-numbered Western Region 31/4s and these were uncommon on the Liverpool–Sheffield services, which were usually in the hands of their northern-based counterparts. I was perhaps half way to having had 31401, when its salvaged power unit hauled me inside the body of 31470 (the former 31161) later in the year! These developments left me with three 31/4s to track down and no idea when further examples would be retired. I also scored 47586 that day when I boarded the 06.33 Poole–Manchester at Macclesfield around lunch time. During the afternoon, a not very well 31461 stuttered into Manchester Piccadilly on the 13.45 Liverpool–Sheffield and no-heat 31221 was duly sent to take the train forward, which was an engine lucky in more ways than one. It had survived a collision two years earlier and unlike other 31s was repaired and returned to service. After then subsequently being withdrawn, it had a second reprieve, when it was reinstated a few weeks earlier and I was in the right place at the right time to score it.

Ten days later (16 April), after spending the morning on some routine moves around Manchester Piccadilly which included 31428, 31461, 47605 and 47602, as I was crossing the viaduct approaching Stockport station behind 47646 on the 14.01 Manchester–Stafford, I saw a Railfreight Grey liveried 31 on the front of a train in the "down fast" platform. This would almost certainly be a required haulage, so I was willing it to wait long enough for me to dash under the subway to reach it. That wasn't to be though and as my train entered the mouth of Stockport station, the other train powered away and I was the recipient of much gloating, shouting and waving from fellow bashers who were on it, who clearly read my deflated expression. This was the 07.50 Norwich–Liverpool, which 31155 had been in charge of from March (the Cambridgeshire town, not the preceding month!). Not wanting to admit defeat, I worked out that the stock would be used for the 14.45 Liverpool–Sheffield, freeing 31155 up from the blocks at Lime Street to work the next hourly 31-hauled train as the diagram dictated. This would be the 15.45 Liverpool–Cleethorpes; however, I knew that a no-heat 31 sticking to the diagram in April was a long shot, with there no doubt being spare traction on Merseyside. I did some fill-in moves, which included a DMU from Stockport to

Below: At the start of the 1980s, push-pull equipment was fitted to a number of the newly-created Class 47/7s, for operation between Glasgow and Edinburgh. 47703 pauses at Haymarket while propelling such a train from Glasgow Queen Street to Edinburgh on 5 April 1988.

Above: 47704 powers away from Haymarket with an afternoon train of matching Mark 3 coaches bound for either Glasgow Queen Street or Aberdeen on 5 April 1988.

Stalybridge, 47452 Stalybridge–Manchester Victoria for 47413 back up Miles Platting Bank on the 15.03 Liverpool–Newcastle and the same unit back to Stockport, where I was poised for the required 31 over the Pennines. When 31461 turned up on the eagerly awaited Cleethorpes service, as I'd already had it that morning and plenty of times before, I lost interest for the day and promptly went home.

The next weekend delivered a formative day in the West Midlands. It was the first of a number of times that I made use of a Day Ranger ticket, the West Midlands equivalent of Greater Manchester's Peak Wayfarer, providing unlimited rail travel for the day within the region. Saturday 23 April 1988 was branded "Midline Day", when West Midlands PTE promoted rail travel by offering thousands of free paper hats and reduced priced tickets (child Day Rangers were a bargain £1.25 instead of the usual £2.00!). Taking advantage of this, a group of us from the North-West took 47543 on the 08.08 Manchester–Poole for the 112 miles to Leamington Spa, which positioned us to pick up the first of the two Class 50 hauled trains to Birmingham (09.17

and 11.17 ex Paddington). The first produced 50031 and dozens of particularly rowdy bashers. As "Hood" was powering up the steep gradient of New Street Tunnel, there was a great atmosphere and it was among the loud declarations of approval that I first heard the word "hellfire" being used to describe the thunderous sound of diesel engine thrash. At New Street, 50031 attracted quite a gathering with bashers encouraging one another to "kiss the bird" on the crest above Hood's nameplate. I didn't fancy that level of intimacy, not least after so many others had already kissed the crest, so I settled for taking some photos and watching it run round. After returning to Leamington behind 47491 (07.45 Newcastle–Poole), we had another storming run back to New Street behind a second winner, 50038. This was such an enjoyable day that we did it all again the following Saturday, when 47487, 50038, 47623 and 47628 turned up on the corresponding services.

At the end of that second Saturday in the Midlands, Paul and I planned to take 85006 back to Manchester on the 17.16 Birmingham International–Manchester; however, the Roarer had failed and the motive power had been upgraded to 31414. We joined this at Wolverhampton and the relatively small 1170 horsepower engine was clearly struggling with low power issues. The weary Goyle limped through Staffordshire at heritage railway speed and we doubted it would make it to Manchester in this condition. Knowing there would be spare Class 20s and 47/3s stabled in the bay platform at the north of Stoke station on a Saturday (this bay platform has since been infilled), there was every chance of a big winner. Not only were we therefore gutted to be told the train was terminating when we arrived into Stoke, but I had a bigger problem. I was going to miss my connection from Manchester Victoria to Hebden Bridge, where I had arranged for my parents to collect and drive me over the moors to Haworth, where they were staying. There was no telephone in the property and this was well before the days of mobile phones, so at Piccadilly I asked the station staff to try phoning their counterparts at Hebden Bridge and pass on a message to my

waiting mother. It was well into the evening by then and Hebden Bridge ticket office was closed. I later learned that while waiting for the train which I wasn't on, my Mum heard the station phone ringing and wondered who it might be, when it was her son desparately trying to pass on a message! When I finally arrived at Hebden Bridge and no one was waiting for me, I made a start on the 20-mile journey over the Yorkshire Moors. I was about half way up the long ascent when the grateful parent and child were reunited and I had much explaining and apologising to do!

The following Saturday was the penultimate weekend of Class 31s working hourly trains in each direction through Manchester, so on 7 May I had a last bash at these and other locos, as I had bigger plans for the following Saturday. I had a decent haul of 31422, 31430, 31441, 31443, 31463, 31464, 47432, 47519, 47553, 47560, 86257 & 87031, but only 47560 (14.01 to Stafford) was new to me, as if to suggest I needed to be travelling further afield.

Above: 50031 has been adorned with a "Midline Day" paper hat and enjoys the adulation of some of the "Roadshow" at Birmingham New Street, after it had entertained many them on the 09.17 London Paddington–Birmingham New Street on 23 April 1988.

Above: 47555 got itself into some trouble and had to be replaced at Stoke-on-Trent on 7 May 1988. A number of weeks beforehand and without any similar drama it was in charge of a parcels train heading for Longsight depot, seen passing through Sheffield on 15 February 1988.

The day ended well, however, and in contrast to the previous Saturday, when the late afternoon Birmingham–Manchester train with an errant diesel was terminated at Stoke. This time, when 47555 failed at Stoke on the 12.40 Poole–Manchester, 47322 was summoned to take the train forward from Stoke, as if to illustrate that good things (or rare rescue locos) come to those that wait.

CHAPTER 6 –
SUMMER SCORES PART I

Saturday 14 May 1988 was a highly memorable day for a number of reasons. It was my 15th birthday, the weather was glorious and the day was spent on what was probably my favourite railtour of all time. It had been some years since a Class 50 had visited Glasgow, so when it was announced that a pair of the class would work from Birmingham and Manchester to Glasgow, Paul and I booked ourselves on "The Hoover Dambuster". It had a comfortably timed mid-morning pick up at Stockport, which gave us the opportunity to fit in two final Class 31 moves on the Liverpool–Sheffield services, before they permanently surrendered to the new 156 Sprinters two days later. 31452 did the honours on the 08.22 Sheffield–Liverpool, followed by 31442 from Oxford Road to Stockport. As 31442 powered away towards the Pennines on the 08.45 Liverpool–Sheffield, there was a sad finality in the air. That was soon forgotten when the compelling

sight of two large logo liveried 50s and the rake of InterCity liveried Mark 1 coaches arrived. 50009 and 50036 took the long way round to Manchester Victoria, via Denton and Ashton Moss Junction, then proceeded to Preston via the "new track" of the Windsor Link; a brand new half-mile stretch between Ordsall Lane Junction and the new Salford Crescent Station. Reliving the days of the locos' youth, the tour then sped up to Glasgow via Shap and Beattock summits.

Being opportunistic or greedy for haulages, depending on how you look at it, we had calculated that if the train arrived in Glasgow on time, we had a "plus two" to make it from the high-level platforms, through the concourse and down to the low-level platforms of Central Station. A Class 303 suburban service could then take us to Dumbarton Central, where we had a similarly tight connection time to meet the 13.00 Oban–Glasgow Queen Street at its

Above: After a highly ambitious dash across Strathclyde and with seconds to spare for a quick photo, 37406 arrives at Dumbarton Central with the 13.00 Oban–Glasgow Queen Street on 14 May 1988.

Above: During an afternoon stop at Barrhead, while waiting for permission to proceed onto the single track section to the south, BR had a relaxed and pragmatic approach, allowing hundreds of enthusiasts to walk trackside and take photographs. 50009 & 50036 would then continue to Birmingham via Kilmarnock, Settle and Manchester on "The Hoover Dambuster" railtour on 14 May 1988.

last intermediate stop. The West Highland Line services to Oban and Fort William were in the hands of Eastfield based Class 37/4s and this high-risk move was the only possible way to fit one in during the tour's afternoon break. All of the Scottish 37s were required locos and so we decided the challenge of this unlikely move was worth attempting. After two very fast sprints at Glasgow Central and under the subway at Dumbarton, this may have been the most ambitious connection that I have ever successfully made. We felt very pleased with ourselves as we headed back to Glasgow behind 37406 and when exiting Queen Street Tunnel, there were large numbers of enthusiasts from the tour on the platforms. We hoped some would recognise us from the tour and wonder how on earth we had pulled off that move! The return journey was via some more scenic and interesting routes. Firstly at Barrhead, hundreds of us poured onto the track for a photostop, before the charter could proceed onto the single-track section beyond. The train continued via Kilmarnock, Settle, Hellifield, Blackburn and Preston, leaving us with

over 500 delightful miles behind the 50s.

The summer timetable period started on 16 May and whilst this marked the end of some haulage opportunities, it brought new possibilities both near and far, as I would soon discover. On the first Summer Saturday, 21 May, I was very pleased to find 37430 on the 05.50 Cardiff–Manchester and 10.00 Manchester–Cardiff return working. This was the first time I had seen a 37 on a passenger train in Manchester and provided me with the first of many Tractor haulages within the region. Other highlights of that day included ticking off my penultimate "Van" (87034), finding no-heat 47288 on the 12.15 Newcastle–Liverpool and a storming afternoon run that left a big impression. A crowd of at least a dozen local bashers shared the day with me, which was fairly typical as one would drift in and out of social groups in line with whoever was out each day. We saw there was a second Cardiff service, comprised of the 14.00 Cardiff–Manchester and 18.17 Manchester–Cardiff, with the inbound working unusually stopping at Alderley Edge. The speculation was that this would be

37 hauled, but in truth I don't think any of us knew, as it was the first Saturday of the new timetable, so we headed down to Alderley Edge on a Class 304 to find out. After working the morning train, 37430 had been swapped for 37426 on the train's second trip, adding a second 37 to my tally for the day. What made the short 13.75 mile run so memorable was the sheer amount of bellowing going on. The coaches were open Mark 2 stock and I remember the view from inside the front coach; someone was kneeling on almost every table, with only their legs and bodies visible, so that the loco could be seen, heard and welcomed to Manchester. There were a few unfortunate normals in that front carriage who were completely bemused to witness this strange behaviour.

Those two return trips from Cardiff to Manchester turned out to be a pretty solid Class 37/4 diagram that summer (although 47458 turned up the following Saturday), as did the Liverpool turn, which consisted of the 05.07 Cardiff–Liverpool, 09.15 Liverpool–Cardiff and the second trip departing Cardiff and

Liverpool at 13.23 and 17.14 respectively. Sometimes, after one of the Cardiff based 37/4s arrived into Manchester, I would ask the driver if I could have a look inside the cab and I don't recall these requests ever being turned down. On one occasion it went a stage further when the driver asked if I would like to ride with him while he ran the loco ran round the coaches. Obviously I accepted and began the exciting experience of my first cab ride by traversing the crossover between Platforms 5 and 6 at Piccadilly, which is probably the "rarest track" I've ever done. After making our way to the station's throat, I vividly remember having to walk through the loco, from one cab to the other, squeezing past the throbbing diesel power unit as it ticked over. I then watched wide-eyed as the driver applied power and effortlessly returned the loco into the station ahead of its return working.

The half term break at the end of May was spent with my family at the cottage in Haworth, although it would be more accurate to say that I used this as a base for the week while I continued chasing locos, doing this for six of the seven days we stayed there. There was plenty of interest around West Yorkshire, with 47s and 31s on the Leeds to Carlisle and Morecambe trains, 47s and the odd 45 on long-distance Cross Country workings and further 47s passing through Leeds on the hourly Liverpool–Newcastle services. Transport was well-subsidised in West Yorkshire and any child day return within the region was a mere 30p. This meant that a 30p return from Keighley to Huddersfield or Wakefield (via Leeds) could buy me four loco haulages. Between 29 May and 1 June, this highly economical method yielded **31108**, 47422, 47439, 47443, 47444, 47450, **47462**, **47521**, **47540**, 47573, 47583, 47585 & **47648** plus a number of HST power cars in and out of Leeds, the bold numbers being the new haulages.

On Thursday 2 June, the theme continued with a family day out from Haworth and although I may have planted the idea, my parents and brothers were very happy to travel by rail if there was an interesting destination or journey to be had. We joined the 08.25 Leeds–Carlisle at Keighley, on which 47648 transported us through Settle, across Ribblehead Viaduct and on to Carlisle. At this point, the Settle

& Carlisle Line was still under the threat of closure by British Rail, even though the prospect of this had led to a rise in passenger numbers and the reopening of seven stations along the route in 1986, all of which had been closed since 1970. Thankfully, its long-term future was assured the following year, when in April 1989 the Government refused British Rail permission to close the route. At Carlisle, while the rest of the family explored the city's attractions, the eldest son was excused in pursuit of some appealing Scottish traction on the former Glasgow & South Western Route via Kilmarnock. A very pleasant 130-mile round trip to Kirkconnel was made, courtesy of Inverness-based 47546 (11.48 Carlisle–Glasgow) and interloping Bristol-based 47620 on the return (10.55 Stranraer–Euston). I really enjoyed this short Scottish sojourn, with the rugged landscape, rural station architecture and highland stag-adorned loco,

Below: The station buildings shown in this 1988 view of Kirkconnel have not survived; however, the semaphore signals and wrought iron footbridge from which the photograph was taken have thankfully made it into 2019. On 2 June 1988, 47620 arrives on the 10.55 Stranraer–London Euston, which it will work to Carlisle and there be relieved by an AC electric locomotive.

all telling me I was travelling through an entirely different region. Old Oak Common's 47573 looked somewhat out of place in Network SouthEast livery at the north-west tip of England and this returned us to Keighley by way of the 16.09 Carlisle–Leeds, after which perhaps it made its way south to home soil.

After dipping my toe in Scotland on the Thursday, I headed to the Midlands on the Saturday to see what could be found in the way of Class 50s. Covering the morning's Cardiff–Manchester train proved worthwhile when I notched up my third of the six Cardiff based 37/4s in the form of 37431. In

Birmingham, it became clear that unlike the previous timetable period, there weren't any Saturday Class 50 diagrams between Paddington and Birmingham, with the exception of one early morning train that I couldn't reach. On Saturdays the type only visited New Street on services to and from the South-West, which as Table 12 shows, would mean visiting Bristol or Exeter, making these long-distance moves a step too far for my budget. I got wind of some very attractive workings in the East Midlands and Wales, however, which diverted my attention for the majority of the remaining Summer Saturdays.

Table 12: 1988 Summer Saturday Class 50 Passenger Diagrams to & from Birmingham New Street

Turn	Service	Class 50 Hauled Section	Dep. New St	Nearest Station Stop	Arr. New St
1	06.00 Paddington–Manchester	Paddington–Coventry	-	Leamington Spa (07.49)	Coventry 08.06
1	08.35 Liverpool–Penzance	Birmingham–Penzance	10.30	Exeter St Davids (13.03)	-
2	07.07 Plymouth–Glasgow & Aberdeen	Plymouth–Birmingham	-	Bristol TM (09.30)	11.29
2	08.50 Glasgow & Edinburgh–Paignton	Birmingham–Paignton	13.58	Bristol TM (15.38)	-
3	08.17 Penzance–Dundee/Glasgow	Penzance–Birmingham	-	Bristol TM (12.45)	14.29
3	11.50 Glasgow–Plymouth	Birmingham–Plymouth	16.48	Bristol Parkway (18.32)	-

Above: My penultimate Peak haulage was on 26 June 1988, when 45113 had little more than a month left in service. The loco, complete with its unofficial name "Athene", is seen leaving Wakefield Westgate, heading south on the Sundays-only 15.35 Leeds–Cardiff Central service.

Above: 31428 started 2 July 1988 in good health, but it was later shut down and didn't operate for its return journey to Sheffield on the 13.10 from Skegness, which was in the hands of railfreight grey liveried 31317. The split-headcode 37 on the left is the standby loco which might have been summoned had this train not been in the hands of a pair of 31s at the start of the day.

My Yorkshire-based bashing friends had been in touch, telling of the abundance of interesting traction that was turning up on the Saturday services to and from the holiday destinations, including Class 20s on the Skegness route. It had been two years since any regular passenger trains had been in the hands of Bombs (as they were then more commonly referred to, above their longstanding Choppers nickname). There were no regular Class 20 passenger diagrams during 1987 and I learned that many bashers feared they had seen the last of these workings. The return of pairs of these fine sounding English Electric engines on weekly trains to Skegness was therefore very welcome, especially to those of us for whom this was a new experience. So on Saturday 11 June, I opened my Class 20 account with 20190 and 20209 on the 10.15 Skegness–Sheffield and then followed the crowd taking the 14.17 Lincoln-bound DMU from Sheffield to Worksop. There we met splitbox 37062, complete with "I love Skegness" sticker on its nose, taking this back to Sheffield on the 13.10 Skegness–Sheffield via Lincoln.

The following weekend (18 & 19 June) was spent around Manchester and yielded 47007 "Stratford" (1320 Manchester–Plymouth) and two new Roarers (85023 & 85035) on the Saturday. I had a generous helping of ten 47s on trains in and out of Piccadilly on the Sunday, as due to the electric power being off during maintenance work, all trains were diesel hauled. The winner to dud ratio was a decent 6:4, with the former consisting of 47663 (09.23 Manchester–Brighton), 47662 (10.26 to Euston), no heat 47330 (11.26 to Euston), 47515 (13.15 to Birmingham), 47551 (12.25 from Birmingham) and 47427 (14.07 to Birmingham). The not required machines were 47608, 47648, 47645 & 47417, completing an excellent day out for Brush engines.

After a weekend at the Yorkshire base, when a couple of 47s around Leeds and 45113 (15.35 Leeds–Cardiff) were squeezed in on 26 June, I was then back in the northern haulage capital of Sheffield on Saturday 2 July. This was a particularly fruitful day and memorable for a number of reasons. My moves are shown in Table 13, which gives an idea of the sheer

Above: With little more than three weeks before its withdrawal when 45110 worked the 09.19 Bristol–York on 2 July 1988, I wasn't to know that this would be my final Class 45 haulage. The Peak is seen attracting the attention of enthusiasts on and off the train as it leaves Sheffield.

volume and variety of loco haulage on offer in one day. I remember the sense of satisfaction as I watched Stratford's 47123 power away from Layton and up the incline towards Blackpool in the last of the day's sunshine, for the final mile of its journey.

Table 13: Moves on Saturday 2 July 1988, a typical 1988 Summer Saturday based around Sheffield

Traction	From	To	Scored	Mileage	Service
DMU	Sale	Oxford Road		5.25	
303	Oxford Road	Manchester P		0.50	
37428	Manchester P	Stockport	Yes	6.00	10.00 Manchester–Cardiff
31409	Stockport	Chesterfield		48.25	09.25 Liverpool–Nottingham Relief
45110	Chesterfield	Sheffield		12.25	09.19 Bristol–York
31309	Sheffield	Chesterfield	Yes	12.25	11.20 York–Cardiff
20056 + 20090	Chesterfield	Sheffield	Yes x2	12.25	10.15 Skegness–Sheffield
DMU	Sheffield	Worksop		15.75	14.17 Sheffield–Lincoln
31317	Worksop	Sheffield	Yes	15.75	13.10 Skegness–Sheffield
DMU	Sheffield	Manchester P		42.75	
47123	Manchester V	Layton	Yes	47.00	19.30 Manchester V–Blackpool N

Seven days later, a contingent from the North-West made our way to what was "Britain's Second City" in terms of haulage that summer – Shrewsbury. The 09.00 Manchester–Euston conveniently connected at Crewe with the (usually) Class 37 hauled 09.15 Liverpool–Cardiff, which probably contained more cranks than normals on Saturdays. This in turn connected with the 07.40 Euston–Pwllheli at Shrewsbury, which a Class 47 handled from Wolverhampton, before a pair of Class 37/0s took it forward on the single track section into Wales. These locos could then be enjoyed for 33.75 miles west to Newtown where there was a passing point, which allowed the train to pass the 10.10 Aberystwyth–Euston. This would be in the hands of two Tinsley-based 37/5s and was usually in Newtown station, ready to depart as soon as the Pwllheli-bound train arrived, necessitating a fast dash across the footbridge (along with dozens of other bashers). Alternatively, one could make the risk-free change at the first station, Welshpool. This gave time to take photographs, although this was at the expense of a further 28 miles of Tractor haulage and involved more than half an hour waiting at Welshpool station. For the hardier bashers, the two 37/5s could be done on their outward journey, the 06.20 Birmingham–Aberystwyth and the 10.10 return working from Aberystwyth. Sombre looking characters who were on this higher mileage move, some of whom had done an overnight in order to be at Birmingham for 06.00, could be seen waiting at Newtown. In contrast to those of us who arrived from Shrewsbury, they would then take the 37/0s forward to Pwllheli and potentially all the way back to Wolverhampton. This gave over 400 miles behind four different no-heat Tractors through some of the country's best scenery, something I would pay good money for the opportunity to do today if paths could be found on the Cambrian Line. For the majority that did the shorter Shrewsbury–Newtown trip, the 37/5s would return

Above: Before Welshpool station was relocated and rebuilt in 1992 to make way for the A483, which is now located where the train above is, 37185 & 37215 are given the road at Welshpool while working the 07.40 Euston–Pwllheli on 9 July 1988.

us to Shrewsbury, where a third pair of 37s could be taken west again on the 09.40 Euston–Aberystwyth, being ETH-fitted examples this time. When reaching Newtown station for a second time, the excitement was punctuated with a 40 minute wait for a Sprinter back to Shrewsbury, which then positioned one for what was usually the eighth Class 37 of the day on the 14.00 Cardiff–Manchester. The moves from my first day out on the Cambrian Line are shown in Table 14 and needless to say, it wasn't the last!

Table 14: Moves on Saturday 9 July 1988, a typical 1988 Summer Saturday based around Shrewsbury.

Traction	From	To	Scored	Mileage	Service
304	Sale	Manchester P		5.75	
86231	Manchester P	Crewe		31.00	09.00 Manchester–Euston
47533	Crewe	Shrewsbury		32.75	10.00 Liverpool–Cardiff
37185 + 37215	Shrewsbury	Welshpool	Yes x2	19.75	07.40 Euston–Pwllheli
37680 + 37380	Welshpool	Shrewsbury	Yes x2	19.75	10.10 Aberystwyth–Euston
37429 + 37428	Shrewsbury	Newtown	37429	33.75	09.40 Euston–Aberystwyth
150	Newtown	Shrewsbury		33.75	13.09 Aberystwyth–Shrewsbury
37431	Shrewsbury	Manchester P		63.75	14.00 Cardiff–Manchester
303	Manchester P	Sale		5.75	

Above: The third pair of Tractors of the day along the Cambrian Route; 37428 & 37429 head west from Newtown while working the 09.40 Euston–Aberystwyth on 9 July 1988.

CHAPTER 7 –
SUMMER SCORES PART 2

The two Saturdays around the Sheffield and Shrewsbury areas in early July were such a success that they provided a template for a number of 1988's remaining summer Saturdays. These were the days when mobile communication devices and computers on which information could be exchanged with those with shared interests were the stuff of science fiction. Reliable gen on which locos were working particular services was therefore pretty hard to come by. A bit of a scene, however, developed by the buffer stops at Manchester Piccadilly at around 07.30 each Saturday morning. I don't know where it came from, but the information shared there at the hour when most teenagers were still some way from even waking up, was never wrong. There tended to be 10 to 20 young men consuming news of the day's big workings and the headlines usually included which 37s were on the Cambrian services, what was working to and from the East Coast resorts and any other interesting locos on passenger trains. With Manchester conveniently positioned for both the East Midlands and Central Wales, some very business-like discussions would then follow, which sharpened the communication, prioritising and budgeting skills ahead of the teenagers' time. The decision-making criteria included who needed which locos, what was affordable and with whom one wanted to spend the day – or avoid! It could require some fast-paced timetable referencing on the platform to see what was possible and most of us carried the full BR National Timetable, or bible as it was known, for that purpose. Next followed an exodus to the ticket office to buy the necessary pieces for the day's travel. At least once, further information came to light as someone new arrived. This required a handful of us to proceed to the Travel Centre at Piccadilly and request refunds and alternative tickets to an entirely different destination! The staff never seemed to mind or ask any questions about this odd behaviour. After settling upon our plans for the day, the crowd would then disperse to various destinations until the same excitable conversation was repeated the following week.

The breakfast-time platform chat on Saturday 16 July revealed that 37215 was working the Pwllheli train for a second consecutive week and had exchanged its partner 37185 for 37062, which a number of us recently had on the Skegness–Sheffield service a few weeks earlier. Whilst the nearby Aberystwyth train was in the hands of the required pair 37679 & 37682, the lower number of winning locos on offer in Wales that day was outdone by what was available in the Eastern Region. 20029 and 20034 were on the "first Skegness–Sheffield" as it was known and the comparatively uninspiring 31443 was on the "second Skegness–Sheffield". We had a difficult decision to make though, as a particularly unusual loco was on offer, working the 10.00 Great Yarmouth–Leeds relief service, which ran on a number of Saturdays that summer. This usually produced an ETH-fitted 31 or 47; however, on this date it was 37689. Unlike its closely numbered Tinsley-based relatives (37676–37688) which could be found on the Saturday Aberystwyth trains, 37689 was a Cardiff-based Tractor and rare for a passenger train, more so outside of Wales. The latest train that could be taken from Sheffield to Wakefield to reach 37689 was the 09.17 Paddington–York, which departed Sheffield at 13.28. Annoyingly this was nine minutes before 20029 & 20034 were due into Sheffield on the 10.15 from Skegness, meaning we had to choose one or the other. There was no way of knowing whether there would be another chance to travel behind loco(s) that were turned down or missed on any given day and the variety, spontaneity and luck at play have always been part of the hobby's fun. A number of us decided on the Cardiff-based 37 above the required 20s and with the hindsight now available, that was probably the right decision, as 37689 worked very few passenger trains after this. Of the 20s turned down that day, I never did track down 20029 before it was withdrawn three years later, but did at least do so for its partner 20034 which I had let go. The other winning engine that day was 47233, which was in charge of the 09.08 Leeds–Yarmouth.

On arriving in Leeds behind 37689, numbers of enthusiasts more akin to those found on a railtour proceeded to pour out of the train and vie for photographs. I was then conveniently positioned to make the short trip across West Yorkshire and be reunited with my family who were spending the weekend in Haworth. Instead of joining them in the car journey back to South Manchester the next day, I opted to make my own way by rail and see what could be found between Leeds, Wakefield and Manchester. Of the five 47s which I travelled behind that Sunday, a useful four were new to me, consisting of 47651 (11.03 Newcastle–Poole), 47424 (12.25 Newcastle–Liverpool), 47423 (13.52 Liverpool–Newcastle) and 47443 (16.25 Newcastle–Liverpool). In addition, instead of the usual HST, I was surprised to find 47503 on the 16.33 Leeds–Kings Cross, so couldn't resist fitting that in too.

Table 15: Examples of the traction found in the Sheffield area on Summer Saturdays during 1988.

Outward Working	Return working	Usual Traction
07.18 Derby–Skegness	10.15 Skegness–Sheffield (via Nottingham)	Usually 2 x 20, or a Class 45 (inc. 45037 on 9-7-88)
09.20 Sheffield–Skegness	13.10 Skegness–Sheffield (via Lincoln)	Anything! 37/0, 1 or 2 x 31, or a no heat 47
Morning Norwich–Yarmouth services	10.00 Yarmouth–Leeds (Relief)	Usually a 31/4 or 47/4, sometimes at 31/1 or 37/0
09.08 Leeds–Yarmouth	15.52 Yarmouth–Leeds	1 or 2 x Class 31 or any 47
06.36 Sheffield–Skegness (via Nottingham)	10.42 Skegness–Leeds (via Lincoln)	Anything! Any 1 or 2 x 31, a 45, a 47/0 or 47/4.
09.31 Leeds– Skegness (via Lincoln)	13.45 Skegness–Leeds (via Lincoln)	Usually a 31, but a 47/0 sometimes

Table 16: Examples of the traction found in the Shrewsbury area on Summer Saturdays during 1988.

Outward working	Return working	Usual Traction
09.15 Liverpool–Cardiff	13.23 Cardiff–Liverpool	37/4
07.40 Euston–Pwllheli	15.05 Pwllheli–Euston	2 x 37/0
06.20 Birmingham–Aberystwyth	10.10 Aberystwyth–Euston	2 x 37/6
09.40 Euston–Aberystwyth	15.25 Aberystwyth–Euston	2 x 37/4
10.00 Manchester–Cardiff	14.00 Cardiff–Manchester	37/4

Once the school term finished in late July, any thought of additional study was far from my mind as I approached the GCSE year and I took the opportunity to step up the loco bashing. Saturday 23 July was a relatively poor day for no-heat traction in the East Midlands, with ETH locos dominating. 45113 worked the "first Sheffield–Skegness" instead of the usual Class 20s, 31459 did the "second Sheffield–Skegness", 47535 was on the 09.08 Leeds–Yarmouth & 15.52 return and 31467 was in charge of the 10.00 Yarmouth–Leeds relief. That may be why I elected not to venture out of Greater Manchester on that day. My notes state that 47356 made a number of trips between Manchester Victoria and Blackpool and despite being within the region, I didn't manage to track it down that day, or indeed ever again before it was converted into the first Class 57 in 1998.

Week one of the holidays was then kicked off with a Coast & Peaks Rover between 25 and 31 July. This was spent in the company of Class 37s on the Cardiff to Liverpool and Manchester services, Class 47s on the Liverpool–Newcastle and Crewe–Holyhead routes and some of the great variety of traction on the Manchester Victoria to Blackpool, Blackburn and Southport trains. The last included a number of 31/4s that had been displaced from the Liverpool–Sheffield route and 31276 on the 17.07 Manchester Victoria–Southport on 29 July. Not wanting to limit myself to the constraints of the North-West and Wales, I made a couple of useful "add-ons" to the Coasts & Peaks Rover, tagging on a West Midlands Day Ranger on Thursday 28 July to capitalise on the weekday Paddington–Birmingham Class 50 working, which began my affinity with 50030 "Repulse". Similarly, a trip along the Cambrian Line to Newtown on the Saturday, resulted in four new NB 37s in the form of 37278 & 37175 on the Pwllheli train and 37682 & 37684 on the Aberystwyth service. The afternoon was then spent traversing the Shrewsbury–Chester line for the first time, to then enjoy runs behind 47359 and 31419 on the North Wales Coast route. The rover was concluded on 31 July, which was another Sunday when engineering work required the overhead power supply to be switched off in the North-West. I consequently benefitted from seven different 47s on Class 1 trains, the highlight of which was the last one of the day. Those caught are listed opposite, with the winning examples in bold.

Above: After another productive Saturday travelling behind no-heat Class 37s on the Cambrian Route, 47359 was found on the North Wales Coast Line during the afternoon of 30 July 1988. It is seen powering away from Flint while in charge of the 11.30 London Euston–Holyhead.

47638	09.23	**Manchester Piccadilly–Brighton**
47440	**10.26**	**Manchester Piccadilly–London Euston**
47413	09.25	Birmingham New Street–Manchester Piccadilly
47411	11.26	Manchester Piccadilly–London Euston
47428	**10.47**	**Birmingham New Street–Glasgow Central**
47487	13.30	London Euston–Manchester Piccadilly
47190	**18.25**	**Manchester Piccadilly–London Euston**

The morning after the Coast & Peaks Rover ended, I made a return trip from Sale to Manchester Piccadilly courtesy of a Class 304 each way. It couldn't have been a trip primarily for loco bashing and was perhaps to browse Manchester's legendary music shops, the likes of which included Power Cuts and Piccadilly Records, as I would do from time to time. I made a single return trip to Stockport and had either been tipped off or had my luckiest break. After taking 86230 to Stockport (not required, so this wasn't the reason for the mystery trip), I then had the extremely rare experience of a pair of Class 20s back to Manchester Piccadilly on a service train, this being the first of only two instances where I managed this. While working the 06.40 Poole–Manchester 47591 had run into trouble in the Stoke-on-Trent area, where it was declared a failure. As 20025 and 20044 were available nearby, they were attached

Above: When 47591 ran into trouble at Stoke-on-Trent on Monday 1 August 1988, 20005 & 20044 were summoned to take the 06.36 Poole–Manchester Piccadilly forward, providing me with one of only two instances of Class 20 haulage into Manchester on a service train. The pair are seen heading out of Piccadilly after being released from the coaching stock that they had brought in.

and dragged the disgraced Type 4 to its destination that Monday afternoon.

The next Saturday was a superb day. In addition to various NB action east of the Pennines, the profitability continued during the evening with 31118, 47229 and 31465 on the western side. The day's moves are shown in Table 17 and as well as five no-heat engines, 31465 is worthy of a mention, with it being my first winning 31/4 for over four months, leaving just two members of the sub-class to track down.

Table 17: A great day for loco bashing on Saturday 6 August 1988

Loco	Move	Mileage	Service
47447	Manchester Piccadilly–Sheffield	42.75	09.08 Liverpool–Yarmouth
156	Sheffield–Chesterfield	12.25	Liverpool–Norwich service
47213	Chesterfield–Sheffield	12.25	09.19 Bristol–York
47591	Sheffield–Chesterfield	12.25	11.20 York–Cardiff
20005 + 20048	Chesterfield–Sheffield	12.25	10.15 Skegness–Sheffield
156409	Sheffield–Bolton	52.25	Nottingham–Blackpool service
31118	Bolton–Manchester Victoria	10.75	Blackpool–Manchester Victoria service
31465	Manchester Victoria–Bolton	10.75	Manchecster Victoria–Blackpool service
150	Bolton–Manchester Victoria	10.75	Unknown
47478	Manchester Victoria–Bolton	10.75	18.20 Manchester Victoria–Glasgow

Above: I rarely photographed electric locomotives as they were so common; however, by 1988 Class 81s on passenger services were rare. When 81010 worked the 16.20 Manchester–Plymouth on 11 August 1988, it turned out to be my final 81 haulage. The train is shown leaving Stockport, with the former bay platform (now infilled) in the foreground and a Class 304 heading for Piccadilly in the background.

Some very useful mid-week moves followed and Monday 8 August demonstrated that a day around Manchester could yield as much as a Summer Saturday around Sheffield, with haulages which included 31118, 31223, 47335, 47571 & 47584 on some of the many loco-hauled trains between Manchester Victoria and Blackpool. The Thursday then yielded new to me 47517 and my penultimate 31/4, 31468 on an afternoon train from Blackpool, as well as 47664, 85036 and my last ever 81 haulage (81010) on trains to and from the city's southern terminus. Examples of Classes 20, 31, 37 and 47 then populated the Saturday's moves, with highlights which included 47203 across the Pennines on the 09.08 Liverpool–Yarmouth and 20060 & 20209 on the 10.15 Skegness–Sheffield. I rounded off that day by making my way to Blackpool, to pause and enjoy the friendship and hospitality of my grandparents, as well as positioning myself for some loco bashing with a different emphasis from that of the preceding weeks.

Above: Class 47/4s with an ample rake of coaches to accommodate the many passengers on cross-country routes (hint to current operators!) were an everyday occurrence during the 1980s. On 3 August 1988, 47654 clears its throat as it accelerates away from Chesterfield with the 15.05 Leeds–Bristol.

CHAPTER 8 –
SUMMER SCORES PART 3

On Saturday 13 August 1988 I began one of my longer stays in Blackpool with my grandparents. My parents were both born in the famous coastal town, spent most of their childhood years there and it's where they met as young adults. My dad's work took the young married couple to south Manchester, where I and my two younger brothers were in turn raised. All four of their parents (my grandparents) retired and remained in Blackpool, where the family has many connections, so regular visits have featured for as long as I can remember. I had a particularly close relationship with my maternal grandparents and many extended stays with them. They enjoyed these as much as I did and as I moved from child to teenager, outings to Gynn Gardens, Stanley Park, the Piers and beach were exchanged for long-distance rail trips together. The three of us loved these days out, as they enjoyed the rail travel and exploring British cities, whereas I enjoyed the journeys and further rail travel at the destination!

I probably arrived a little bedraggled, with the smell of diesel hanging on my clothes, following my close encounters with members of Classes 20, 31, 37 and 47 earlier in the day. We spent the Sunday together, which would have involved going to church, followed by what they called a "snifter" (a glass of sherry preceding a meal) and a roast dinner of the highest quality. I then spent the Monday pottering about around Preston and Warrington and managed to travel behind six different 47s on the Preston–Blackpool route, the highlights being scratching 47051 and 47625 on Blackpool–Manchester services. We were up early the next morning for the first of our long-distance trips, starting with 31223 to Preston on the 07.28 Blackpool–Manchester Victoria. We proceeded to Glasgow Central behind 85020 and then parted company to pursue our different interests. I walked to Queen Street station and then spent four wonderful hours in the company of some of the most beautiful scenery to be seen from a carriage window and two new Scottish 37/4s on the West Highland Line. 37405 took me north to Crianlarich (12.04 Glasgow–Oban), where it was timetabled to pass the corresponding southbound service. This was a risk-free four-minute connection, as the two trains have to wait for each other at Crianlarich before proceeding along the single track sections north and south. The

station has an island platform, meaning no footbridge or subway dash is required and there is time for a quick photo. 37410 returned me to Glasgow (12.50 Oban–Glasgow) and I spent the latter stages of the journey in the carriage vestibule immediately behind the engine, to savour as much of the engine's sight and sound as possible. The evening journey back to England involved 86250 to Preston and the newly created 31400 forward to Layton.

For the following week's day out, the three travellers started the day with 86405 along the other half of the West Coast Main Line to London Euston. The capital always had a special place in my Gran and Grandad's hearts, as they met there in the most dramatic of circumstances. During VE Day in May 1945, many thousands of people were celebrating in Trafalgar Square and among the masses two northerners found themselves in close proximity. The local government clerk and nurse had both relocated to London during the war and were in their mid-20s at that point. News that the King (George VI) and Prime Minister Winston Churchill were to appear on the balcony of Buckingham Palace started to filter through the crowd, which began to head in that direction. As they jostled along The Mall towards the palace, the helpful man took care of the nearby lady who was a couple of inches below the average head height. Their friendship began that afternoon and time demonstrated that it wasn't just founded upon the drama of the occasion's celebrations. Well over 50 years of marriage followed, making VE Day significant for them in more ways than one. Returning to central London with their grandson approaching 40 years later, they perhaps reflected on that first encounter. Alternatively they may have compared the city's sights to those they had seen in Glasgow the previous week, while I contrasted the West Highland Line's English Electric Type 3s with the Great Western Main Line's English Electric Type 4s. I made the most of a £4.20 child day return from Paddington to Oxford, by breaking the journey into four sections, with two short breaks in Reading, so I could enjoy four different Class 50s that afternoon. Two of the 50s were new to me that day, edging my tally up to 11; however, when sitting behind such fine machines, listening to their engines hammer through the likes of suburban London and Sonning

Above: Crianlarich, roughly half way between Glasgow and Fort William, was the destination on 16 August 1988 and provided a convenient location to transfer between the loco-hauled trains which pass there. After 37405 has just arrived and sits behind the camera, 37410 meets its classmate with a long rake of Mark I coaches on the 12.50 Oban–Glasgow Queen Street.

Cutting at 100mph, whether they were required or not didn't seem to matter.

On arriving back into London shortly after 17.00, I would have loved to stay and watch the rush hour comings and goings at Paddington and choose another loco to head west behind. There was ample choice, with 15 different loco diagrams (11 x Class 50s and 4 x Class 47s), plus 6 HSTs in or out of Paddington in less than two hours each weekday. We had beds to reach 224 miles away though, so 86261 raced us up the West Coast Main Line to Preston (18.30 Euston–Blackpool), where 47521 then replaced it at the front of the Mark 3 coaches for the final 17.5 miles to Blackpool.

Table 18: Class 50s enjoyed on 25 August 1988:

Loco	Travelled From & To	Service
50012	Paddington–Reading	13.05 Paddington–Newbury
50024	Reading–Oxford	13.15 Paddington–Oxford
50033	Oxford–Reading	16.00 Oxford–Paddington
50030	Reading–Paddington	14.41 Birmingham–Paddington

There were eight days between our trips from Blackpool to Crianlarich and London. The first seven of these were spent doing a North-West Rover and one was then used to rest from travelling. This rover on more familiar territory gave unlimited travel from

Manchester, Liverpool and parts of West Yorkshire in the south, up to Carlisle including the Settle & Carlisle Line. For the diesel basher, there was plenty to choose from and the main routes I capitalised on were Manchester–Blackpool and three of the routes radiating from Carlisle (to Leeds, Newcastle & Glasgow). I found dozens of Class 47s and 31s on these, including various no-heat examples and for most services, the locos in charge changed every day. On four of the seven days, I made my way up to Carlisle with Andrew from Bamber Bridge who was one of the the regular local faces and also equipped with a North-West Rover that week. We discovered a particularly productive move, which started with the 07.28 Blackpool–Manchester, a train that was usually loco-hauled and could be in the hands of a 31, 47 or a Sprinter. At Preston this connected with the 07.10 Manchester–Glasgow & Edinburgh forward to Carlisle, where we could see what the 06.55 Berwick–Carlisle produced. As the normals alighted this train from the North-East, we would discretely enter the Mark 1 coaches and hide in a toilet. This was because it

terminated in one of Carlisle's southern bay platforms (5 or 6) and in order to free up the incoming loco for the return working, a shunter would haul the stock out of the station towards London Road Junction. Apart from enjoying the cheeky challenge of not being caught by the guard or station staff, this gave us some rare Class 08 haulage. At the time Carlisle Kingmoor had an allocation of 13 Class 08s, which could be seen on various shunting duties around Carlisle. As Table 19 shows, on the four days we did this move, we benefited from three different 08 haulages. After releasing the loco that had arrived from Berwick, the shunter would return the stock into Citadel station and the main line engine would be reattached, to form the 10.23 Carlisle–Newcastle. This could then be taken across the Tyne Valley Line to Bardon Mill, giving a +13 connection for the loco-hauled 10.30 Newcastle–Carlisle to its destination. There was the higher risk option of continuing to Haydon Bridge, for a +2 and we must have felt confident enough to risk this on the fourth occasion, as the connection was successfully made.

Below: 47236 was one of many members of the class to work on the Settle & Carlisle line that particular week and is shown leaving Lazonby & Kirkoswald with the 12.42 Carlisle–Leeds on 20 August 1988.

Above: Even with just a two-minute connection at Haydon Bridge, there was still time to photograph 31286 after alighting from the short train which formed the 10.23 Carlisle–Newcastle, before 47455 would return us to Carlisle on 23 August 1988.

On one of the days we crossed from the eastbound to the westbound platform at Bardon Mill, we were surprised to be accompanied by an elderly American gentleman, who also awaited the return working. Both parties were a little sheepish and felt it necessary to explain why we had made a return journey to the small rural station without leaving the platforms. It turned out the need to lower the droplight window to access the external door handle had confounded him to the point that he missed his stop at Haltwhistle, where he intended to alight and explore nearby Hadrian's Wall. As a result, he had travelled the additional five miles unintentionally. We therefore provided him with a slam-door egress demonstration and trust that he found his way on to Hadrian's Wall that day.

The next leg of this productive series of moves was to see what had arrived at Carlisle on the 08.25 from Leeds while we had meandered alongside the upper reaches of the River Tyne. The loco was booked to return to Leeds on the 12.42 departure, which gave time for a leg stretch and lunch break in Carlisle. We did this to Lazonby & Kirkoswald on each of the four days, as the next train from Leeds could

then be taken back to Carlisle and I recall quite a number of enthusiasts did this 30-mile round trip. There was then the option of another two Class 47s on the Carlisle–Glasgow via Kilmarnock route, something I only did on 18 August, courtesy of 47564 and 47432 on the 14.00 Carlisle–Glasgow and 10.55 Stranraer–Euston respectively. As Table 19 shows, a good variety of traction was available, including the unique 31101 which seemed to follow me during the late 1980s. It is one of the two pilot scheme class members that survive, the other being 31018 which is on static display at the National Railway Museum, York. Following collision damage in 1967, the loco to become 31101 was rebuilt and fitted with electro-pneumatic control, becoming an additional member of the 31/1 production sub-class. This meant that it evaded withdrawal, as the other non-standard pilot scheme locos had all been withdrawn by the early 1980s. Being the oldest example, it was given the first TOPS number in the series and is now one of the oldest extant main line diesel locos. Now in its seventh decade of service, it can be found earning its keep on the Avon Valley Railway.

Table 19: Haulages around Carlisle between 18 and 23 August 1988.

Date	08 on the stock shunt	10.23 Carlisle–Newcastle	10.30 Newcastle–Carlisle	10.42 Carlisle–Leeds	10.45 Leeds–Carlisle
18 Aug 88	08690	47522	31101	47553	47427
20 Aug 88	08911	31101	47455	47236	47427
22 Aug 88	08768	47207	143 Pacer	47482	47438
23 Aug 88	Unknown	31286	47455	47438	47622

This week-long North-West Rover also involved a mixture of routine and more interesting haulages. The former included 31406, 31417, 31426, 31431, 31468, 47424, 47435, 47442, 47449, 47459, 47490, 47536, 47592, 47645, 85013, 85025, 85030, 86230, 86254, 86418 and 87010. Other than those in the table above, other notable haulages that week were 31309, 47108, 47196 & 47286 on Blackpool to Manchester or Euston trains, as well as my first ride behind one of the brand-new Class 90s, 90004 being my first.

Despite arriving into Blackpool North little more than eight hours earlier, my accommodating grandparents allowed me to be back there on Friday 27 August in time for the 07.28 departure to Manchester Victoria. The day consisted of a transfer between two of my satellite bases, travelling from Blackpool to Haworth and I assume the early start was to align my movements with the times of diesel-hauled trains. 31305 took me from Blackpool to Bolton, for 47621 on to Manchester Victoria, where I changed for 47424 over the Pennines to Leeds. From there 31461 took me to Keighley, on a Leeds to Morecambe or Carlisle service. After not having seen my parents or brothers for more than

Below: The popularity of Class 20s on passenger trains is evident from the numbers of observers on the platform at Sheffield. After 20042 and 20034 had worked the 10.15 from Skegness, the pair are about to take the empty coaching stock to Derby on Saturday 27 August 1988.

two weeks, it seems that I again only had a brief pit stop, as I was back in Leeds for breakfast time the next morning. As it was a Summer Saturday, I didn't want to miss anything and kicked off the day by taking 31448 south on the 09.31 Leeds–Skegness and made my way to Sheffield to see what else I could find. The highlight of the day was a winning pair of 20s, 20034 & 20042, the first of which was one of the two I had turned down a few weeks earlier in favour of 37689. I managed to miss 31155 and 31166 on the 10.42 Skegness–Leeds, which was extended to Newcastle that day, two locos that I didn't subsequently manage to track down. Had I been on my home side of the Pennines, I would have found the rare occurrences of 37084 deputising for a Sprinter on the 08.10 Cardiff–Manchester and 12.12 return working and 37372 on the 14.00 Cardiff–Manchester and 18.17 back to Wales. I learned a long time ago though, that there's no point in rueing what one has missed out on, but instead to appreciate what one has had – a lesson for bashing and for life in general too.

The following four weekends in September and the first weekend in October included the last five Summer Saturdays of the year.. I made the most of these with further lucrative trips to the East Midlands and Wales on the Saturdays, along with a number of Sunday and weekday outings. The number of loco haulages involved are too large to list; however, some of the highlights are shown in Tables 20 and 21, including all my moves from the weekend of 17–18 September, which was particularly

productive. This period also included one of my all-time favourite bashing days, when on 24 September 1988 I had 37015 & 37174 for the 237-mile round trip from Shrewsbury to Pwllheli. There were plenty of other bashers on this, including a contingent of my regular companions from the North-West. Other than the fantastic scenery of Central and Western Wales, where the single track navigates estuaries and clings to the coastline, this day was memorable for the masses of clag that 37015 produced. With every application of power, the thrash was accompanied by the photogenic sight of thick black emissions. One lad from Wigan who I will protect the anonymity of, had a plastic bag and spent quite some time leaning out of the front window, unsuccessfully attempting to catch a sample of this in his "clag bag"! On the return journey, the train was timetabled for a lengthy stop at Machynlleth, giving time for the bashers on board to perform seminars. In line with this tradition, dozens of bashers arranged themselves at the side, front or even on the locos themselves, to pose for photos. This included people standing on the track, while the driver and secondman remained in the cab and didn't bat an eyelid. By the following Saturday, a number of us had had our photographs from the Pwllheli trip developed and an impromptu "37015 Clag Photo Comparison" session developed on the platform at Stockport. Surprisingly few of these seem to have been shared online since and if any readers have photos from this day, including those of the seminars at Machynlleth, I would be very interested to hear from them.

Table 20: Moves on 17 & 18 September 1988.

Date	Traction	From	To	Score	Service
17 Sep 1988	85023	Manchester Piccadilly	Stockport		09.00 Manchester–Euston
17 Sep 1988	37431	Stockport	Manchester Piccadilly		05.50 Cardiff–Manchester
17 Sep 1988	47348	Manchester Piccadilly	Sheffield	Yes	09.25 Liverpool–Nottingham
17 Sep 1988	47312	Sheffield	Chesterfield	Yes	09.08 Leeds–Yarmouth
17 Sep 1988	31419+31464	Chesterfield	Sheffield		09.19 Bristol–York
17 Sep 1988	31465	Sheffield	Chesterfield		Unknown
17 Sep 1988	20005+20048	Chesterfield	Sheffield		10.15 Skegness–Sheffield
17 Sep 1988	DMU	Sheffield	Worksop		
17 Sep 1988	37072	Worksop	Sheffield	Yes	13.10 Skegness–Sheffield
17 Sep 1988	DMU	Sheffield	Stockport		
17 Sep 1988	37431	Stockport	Manchester Piccadilly		14.00 Cardiff–Manchester
18 Sep 1988	DMU	Manchester Victoria	Bolton		

Date	Traction	From	To	Score	Service
18 Sep 1988	31191	Bolton	Manchester Victoria		13.36 Blackpool–Victoria
18 Sep 1988	47306	Manchester Victoria	Bolton	Yes	15.10 Victoria–Blackpool
18 Sep 1988	31421	Bolton	Manchester Victoria		14.28 Blackpool–Victoria
18 Sep 1988	31191	Manchester Victoria	Bolton		16.10 Victoria–Blackpool
18 Sep 1988	DMU	Bolton	Manchester Victoria		

Table 21: Other notable moves during the last month of the Summer 1988 Timetable Period.

Date	Traction	From	To	Score	Service
3 Sep 1988	37096+37372	Shrewsbury	Newtown	Both	07.40 Euston–Pwllheli
3 Sep 1988	37683+37380	Newtown	Shrewsbury	37683	10.10 Aberystwyth–Euston
3 Sep 1988	47358	Manchester Victoria	Leeds	Yes	13.22 Holyhead–Hull
10 Sep 1988	37101+37174	Shrewsbury	Newtown	Both	07.40 Euston–Pwllheli
10 Sep 1988	37686+37679	Newtown	Telford	Both	10.10 Aberystwyth–Euston
14 Sep 1988	31222	Bolton	Manchester Victoria	Yes	Unknown
14 Sep 1988	31191	Manchester Victoria	Bolton	Yes	14.20 Victoria–Blackpool
24 Sep 1988	37015+37174	Shrewsbury	Pwllheli	37015	07.40 Euston–Pwllheli
24 Sep 1988	37174+37015	Pwllheli	Shrewsbury		15.05 Pwllheli–Euston
1 Oct 1988	20063+20032	Nottingham	Sheffield	Both	10.15 Skegness–Sheffield

Below: 37174 and 37015 were in charge of the Saturdays only 07.40 Euston–Pwllheli on 24 September 1988 and in addition to the beautiful Welsh scenery, the clag that 37015 consistently produced all day was the other memorable sight from the day.

CHAPTER 9 –
WINTER WANDERING

As the summer turned into autumn, my bashing endeavours eased off slightly; this could have been for a combination of reasons. Having recently begun what is now known as Year 11, the talk of this being an important year and the need to apply oneself for the looming GCSE exams may have redirected my attention, although I wasn't renown for my application to school work! Alternatively, it could have been that I had run out of money after all the summer travelling, the hobby still largely being funded by two weekly paper rounds. Also, in October 1988, Manchester Piccadilly was closed for around two weeks while the layout of the track approaching the station was remodelled. During this period some loco-hauled trains started and terminated at Stockport, while others proceeded to and from Manchester Victoria via Denton. Some local stopping services continued

as far as a temporary platform that had been built opposite Longsight depot, giving the opportunity of a rare move to Longsight, where trains connected with a bus to Piccadilly. I remember concealing my pleasure as I stepped onto the little platform opposite the depot buildings, while fellow passengers made their irritation at this inconvenience known to the BR staff – rail replacement buses have been around for as long as I can remember. There were some interesting haulages to be found during this time, such as when I had 47337 out of Manchester Victoria on the 10.30 to London Euston on 8 October. As temperatures fell through October, however, instances of no-heat locos on passenger trains became more uncommon. They weren't unheard of, though, and 47186 was one of the first locos I had over Manchester Piccadilly's new track layout on Sunday 16 October, followed

Below: This busy scene at Stockport is likely to have arisen from the Euston-bound train originating at Manchester Victoria while Piccadilly station was closed for remodelling, leading to more passengers boarding here. No heat traction is employed on 8 October 1988 when 47337 heads south on the 10.30 Manchester Victoria–London Euston.

by 47258 on both the 16.20 Manchester Victoria–Blackpool and 17.58 return working six days later.

The winter timetable period saw the return of a Saturday diagram that brought a Class 50 from Paddington to Birmingham and back. By then the class was a firm favourite of mine, along with a number of my regular travelling companions, including Paul from school and a friend from Sheffield. We made 12 trips to the Midlands during the final months of 1988 and early 1989, as this was the best way to reach the class, other than a much longer and costly day to the capital. The usual move was the Class 47 hauled 08.08 Manchester–Poole as far as Leamington Spa, for a 50 back to Birmingham and a reverse of this during the afternoon. Members of the class encountered during this period were 50025, 50031, 50032, 50039 and 50046, with all except 50032 being found on this working on two or more occasions. Table 22 shows the traction on the Saturdays through the winter period; it wasn't always a Class 50.

Table 22: Locos that worked the 09.40 Paddington–Birmingham & 14.41 Birmingham–Paddington on Saturdays during the winter of 1988–89.

Date	Loco	Date	Loco	Date	Loco
05-Nov-88	50046	07-Jan-89	50031	11-Mar-89	50039
12-Nov-88	47573	14-Jan-89	50035	18-Mar-89	50035
19-Nov-88	50046	21-Jan-89	50030	25-Mar-89	50050
26-Nov-88	50025	28-Jan-89	50031	01-Apr-89	50039
03-Dec-88	50046	04-Feb-89	50037	08-Apr-89	50050
10-Dec-88	50039	11-Feb-89	50046	15-Apr-89	50023
17-Dec-88	50030	18-Feb-89	50031	22-Apr-89	50034
24-Dec-88	50039	25-Feb-89	50039	29-Apr-89	31442
31-Dec-88	50035	04-Mar-89	50032		

Below: With the last of the Class 25s having recently been withdrawn from the main line, the young diesel preservation movement was ripe for expansion in 1988. The Keighley & Worth Valley Railway held an early diesel gala, during which D5054 (24054) & D5209 (25059) are seen at Oxenhope after working in multiple from Keighley on 5 November 1988.

Above: Before passenger trains returned to the Stockport–Altrincham route in 1990, it was used only by diesel-hauled freight trains. This provided sufficient incentive to occasionally draw me to the site of the former Baguley station, to which I could cycle in less than half an hour. Pairs of Class 37s were the most common traction on the limestone trains between Great Rocks Quarry in Derbyshire and the ICI Plant in Northwich; however, on 22 November 1988 I found 47531 in charge of such a working heading for Northwich.

The weekend of 5–6 November 1988 introduced an entirely new concept to me and one that today's fans of diesel locos would probably be lost without – a diesel gala at a heritage railway. The Keighley and Worth Valley Railway (K&WVR) is a preservation pioneer in a number of ways; it's one of Britain's earliest heritage railways, having been operating trains since 1968, and one of the first to realise the potential of diesel galas. The K&WVR had recently acquired its first diesel loco, Class 25 No. D5209, which had recently carried out its first working on the line in October 1987. I recall a conversation with one of the railway's workers at that time, who explained how much easier the diesel was to start up, in comparison to its steam counterparts which needed hours of preparation (which no doubt was the experience of BR's drivers during the 1950s and 1960s as diesel replaced steam). Whilst plenty of diesels could still be found on the main line in 1988, popular classes such as Rats (Class 25), Deltics (Class

55), Westerns (Class 52) and Whistlers (Class 40) had all been withdrawn and the diesel preservation movement was beginning to gain momentum. Along with many from the bashing world, I was intrigued by this event, which offered an inaugural run behind some of these "extinct" classes. I also felt a degree of ownership for the K&WVR, as my mum was a volunteer at Oakworth and Keighley stations and I had watched countless trains pass along the line from the living room in our little holiday house in Haworth. Five locos were involved when I sampled the gala on the Saturday; the railway's resident Rat was joined by Class 24 No. D5054, which provided a striking BR green liveried pair of Sulzer engines. The other visitors were the very different looking and sounding D1041 "WESTERN PRINCE", 55015 "TULYAR" and 55016 "GORDON HIGHLANDER". On the Sunday, the diesel locos shared the line with GWR 2-6-2T No. 6106 and one of the railway's long-term resident Waggon & Maschienenbau diesel

Table 23: New Haulages by Class, 1986–1988

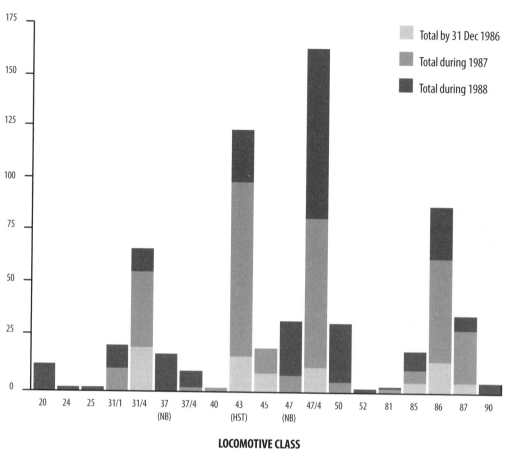

Legend:
- Total by 31 Dec 1986
- Total during 1987
- Total during 1988

LOCOMOTIVE CLASS

railbuses. With there still being plenty of diesels to occupy me on the main line, five years would pass before another diesel event would entice me to visit a heritage railway.

Late 1988 saw railways become a major talking point for the wrong reasons, when on 12 December a commuter-filled electric multiple unit ploughed into the rear of another stationary unit, south of Clapham Junction. A third passing train then collided with these and the incident, which was the result of a recent signal wiring fault, claimed 35 lives and injured a further 484 people. Whilst saddened by such bad news, this didn't quell my appetite for rail travel and I rounded off the year with visits to the Midlands, primarily in pursuit of Class 50s. During that time I also had a few of the new Class 90s that were starting to appear on West Coast Main Line services. Much of my travels during 1988 have been

outlined above, although there were plenty more routine journeys and days out that haven't been mentioned. As well as in words, a basher's activities can also be measured in numbers and the graph above shows the variety of traction I travelled behind in the late 1980s, with all but the five locos at the K&WVR Diesel Gala and four on railtours being found on main line service trains.

Towards the end of 1988, comments on the social network (conversations on trains and on platforms) revealed there were some very desirable haulages on offer during planned engineering works. On Saturday nights and Sundays the West Coast Main Line (WCML) was closed between Preston and Wigan North Western, during which trains were diverted along the non-electrified line towards Manchester, as far as Lostock Junction, where they reversed and headed to Wigan via

Ince. For the Preston–Lostock section, ETH-fitted Class 47s were used and instead of these running round the stock and then doing so again at Wigan, alternative traction was attached to the rear at Lostock. This dragged the train forward to Wigan, where the 47 was then facing in the right direction to continue south. With the 47 covering the need for train heating during the winter months, the no-heat freight locos stabled at nearby Wigan Springs Branch depot could be used for the 6.75-mile section between Lostock and Wigan. These workings were known as the "Lostock Drags" and involved two loco diagrams, one was booked for a Class 47 and the other a pair of "Speedlink Freight General Pool" Class 20s; however, both were usually worked by pairs of 20s. I learned that these diversions had operated during a number of previous years during

the 1980s; for this timetable period they ran on five Sundays in late 1988 and the first six Sundays of 1989. I wasn't to know at the time, but this series of workings would be the last to utilise Class 20s on diverted WCML passenger trains.

Eight trains were involved, four during the early hours of Sunday mornings – all of which were inaccessible for me – and four during the daytime, which were still tricky to access from Manchester. All but one train travelled south, with the locos running light engine from Wigan to Lostock between duties. The diagrams are shown in Table 24. Passenger trains between Bolton, Lostock and Wigan usually travelled to nearby Wallgate station and as only one track connects this route with Wigan North Western, the north and southbound trains passed en-route during the afternoon and didn't connect at Wigan.

Table 24: 1988/89 Lostock Drag Loco Diagrams
Turn 1 – Class 47 diagram which usually produced Class 20 x 2

Service	Departs	Dragged Section	Arrives
22.50 Glasgow–Euston	02.47 Preston	03.27 Lostock–Wigan	
19.35 Inverness–Euston	04.45 Preston	05.22 Lostock–Wigan	
08.24 Blackpool–Euston	08.51 Preston	09.27 Lostock–Wigan	09.59 Wigan
12.25 Preston–Bristol	12.25 Preston	13.12 Lostock–Wigan	13.34 Wigan
10.00 Euston–Preston	14.15 Wigan	14.15 Wigan–Lostock	15.21 Preston

Turn 2 – Class 20 x 2

Service	Departs		Arrives
22.40 Edinburgh–Euston	03.04 Preston	03.44 Lostock–Wigan	
20.45 Aberdeen–Euston	05.23 Preston	06.00 Lostock–Wigan	
13.30 Preston–Euston	13.30 Preston	14.14 Lostock–Wigan	14.37 Wigan
Note: 10.00 Euston–Preston & 13.30 Preston–Euston pass each other between Wigan and Lostock			

The line from Manchester to Bolton was also closed for engineering works on Sunday mornings, with replacement buses between Victoria and Bolton stations. The timings of these meant it wasn't possible to reach Preston in time to be on the 08.24 Blackpool–Euston, which was the only way to travel behind both pairs of 20s. This was because the locos on Turn 2 only worked the 13.30 Preston–Euston, which followed the 12.25 departure from Preston and during its 67-minute journey from Preston to Wigan, this passed the train on which the other 20s were operating. In other words, unless you could be at Preston before 08.50 on Sunday, you had to choose one of the two pairs of 20s. Fortunately for me, I had

very sympathetic grandparents in Blackpool who I could stay with and so 1989's activities began with 50031 to Banbury on Saturday 7 January and a late evening arrival in Blackpool. My Gran was away that weekend and when I asked my Grandad if he could give me a lift to Blackpool North for 08.00 the next morning, a matter of hours after I arrived, he agreed but his reaction revealed that I was pushing my luck with his hospitality! I went ahead with my plans and the Sunday provided me with three new Class 20 haulages (20120, 20113 & 20197). For the following two Sundays I endured a journey comprising of a bus into Manchester, a rail-replacement bus forward to Bolton and then a DMU to Preston. This was rewarded with 20117 & 20065 and 20051 & 20055, the pairs each working the 12.25 Preston–Bristol seven days apart. I chose this train above the second southbound service, as it gave the option of a second run behind the English Electric engines on the northbound 10.00 Euston–Preston. Events on the second of these dates, however (22 January), left me wishing I had waited 65 minutes for the 13.30 departure, because unbeknown to me, the 10.00 Euston–Preston was running very late. As this was decades before anyone had dreamed up the concept of websites where trains' progress could be monitored, let alone portable devices on which to do so, I was oblivious to the fact that I could have had 20147 & 20160 to Wigan on the 13.30 and made the usually impossible minus 22 minute connection for 20051 & 20055. While I was waiting at Wigan for a second run behind the latter pair, to say I was annoyed when the required 20147 & 20160 rolled in on a train that I had turned down earlier, would be something of an understatement. I was angry because I felt I should have chosen the second train from Preston, just in case the northbound working

was late – as it turned out to be, in which case I could have had both pairs. Sadly 20147 was withdrawn within four months and 20160 met the same fate the following year. Both remain unhighlighted in my copy of Diesel & Electric Loco Register.

On Saturday 28 January, the activities of three weeks earlier were repeated, with a long day in the company of 50031 on the Paddington–Birmingham service, after which I made my way up to Blackpool where I arrived during the evening. Perhaps feeling sheepish about taking liberties with my grandparents' hospitality three weeks earlier, instead of another early start and day of Bomb bashing, I spent some time with my relatives and made my way back to Manchester later on the Sunday. As it happened, I needed all four of the 20s that I turned down that day (20019 & 20057 and 20028 & 20172) and despite having dozens of further Type Ones in the years that followed, I didn't have another opportunity to catch up with any of those four. I made sure I didn't miss out on the final two Sundays of Class 20 action though,

Above: The highly unusual combination of Type 1 locomotives built as early as the 1950s hauling a West Coast Main Line passenger train consisting of Mark 3 coaches. 20043 & 20069 arrive into Wigan North Western station, having hauled the 13.35 Preston–London Euston from Lostock Junction on Sunday 12 February 1989, before 47457 takes the train south.

and as Table 25 shows, there were three and four different pairs in operation on 5 and 12 February respectively. On the first of those days I endured the replacement bus ordeal in order to reach 20100 & 20173 and rounded the series of Sundays off with another 08.24 start from Blackpool North,

to position myself to benefit from four new Class 20 haulages. It was an enjoyable and at times frustrating stint, but overall very productive, yielding 13 new Class 20s over the five Sundays and a particular appreciation of the sound of their engines, which remains today.

Table 25: Locos working the Lostock Drags during engineering work on 11 Sundays during WCML closures.

Date	Turn 1	Turn 2	Additional Locos (usually on overnight trains)
20 Nov 1988	20219 + 20224	47497	
27 Nov 1988	47125	20051 + 20055	
4 Dec 1988	20100 + 20173	20052 + 20058	20010 + 20175
11 Dec 1988	20189 + 20218	20088 + 20187	
18 Dec 1988	20080 + 20135	20088 + 20187	
8 Jan 1989	20090 + 20120	20010 + 20175	20113 + 20197
15 Jan 1989	20065 + 20117	20090 + 20120	
22 Jan 1989	20051 + 20055	20147 + 20160	
29 Jan 1989	20019 + 20057	20028 + 20172	
5 Feb 1989	20028 + 20172	20007 + 20053	20100 + 20173
12 Feb 1989	20004 + 20052	20043 + 20069	20028 + 20172 & 20147 + 20160

One other loco-hauled run from the final weekend of the Lostock Drags is worthy of a mention. My journey to Blackpool on the Saturday included the 15.45 Manchester Victoria–Glasgow & Edinburgh as far as Preston. This service was part of a Class 47 diagram that consisted of three return trips from Manchester to Preston, finishing with a run to Blackpool on Saturdays and to Barrow in Furness on weekdays. Unusually, though, on Saturday 11 February, the Scotland-bound train was in the hands of 31408. As this consisted of eight or nine coaches, the 31 had to work a bit harder than a 47 would need to, with it having an

engine less than half the size. Consequently, and as I suspect the driver was in a rush that afternoon, this was the loudest and most memorable run I've ever had on a 31. I spent much of the journey at the front window of this train, sharing it with a Lancashire-based basher and together we marvelled at the noise it was making. Much as I enjoy chipping away at new haulages, sometimes you just have to savour the moment while the changing landscapes pass you by. Those 31 miles from Manchester to Preston packed in at least twice the usual thrash, which I haven't forgotten 30 years later!

CHAPTER 10 –
SOME SPECIALISATION

After the sun set on the final day of the "Lostock Drags", the next morning marked the start of the week-long February half term school break. Back to my unrelenting pace of travel, I was up early the next morning (13 February) and back in the city centre in time to take 87031 on the 08.00 "Manchester Pullman" from Piccadilly to Stoke-on-Trent. This gave me 12 minutes to buy the necessary ticket before I continued to Leamington Spa behind 47659 on the 08.08 Manchester–Poole. Having made good use of the four other Rail Rovers that were accessible from Manchester during the last 18 months (Coast & Peaks, North-West, North-East and East Midlands), it was now time to try out a Heart of England Rover. This ticket is still available today and has a slightly larger area of validity than that of 30 years ago, with unlimited travel (after 09.00 Mon–Fri) from Chester, Crewe, Stoke and Derby in the north, to

Northampton, Oxford, Gloucester and Hereford in the south. One of the main appeals of this region was access to Class 50s on the daily services they worked from Paddington to Birmingham and to Wolverhampton on Saturdays which provided some further mileage. Annoyingly, the rover was valid from Birmingham to Banbury and also to Oxford, but only via the longer route through Worcester and Evesham, which in 1989 only saw occasional Class 50 hauled trains (the only 50 diagrams between Paddington and Hereford ran late on Sunday afternoons, making these too difficult for me to cover).

This rover seemed to centre on 50025, with runs behind "Invincible" every day that I travelled. On 13, 15 and 16 February it was in charge of the 09.40 Paddington–Birmingham and 14.41 return working; on Valentine's Day it was given a change of scenery, appearing on one of the Berks & Hants

Above: When this photograph was taken on 14 February 1989, 50040 carried the "Centurion" nameplates, which had been transferred from 50011 shortly after its withdrawal in 1987. The loco arrives into Oxford before working the 15.00 to London Paddington.

Above: 37407 continued to carry the name and West Highland Terrier transfer that marked its former Scottish identity for some time after it relocated to Wales. A few months after providing a winning haulage on the 10.00 Manchester Piccadilly–Cardiff Central, it passes Coedkernew, west of Newport, while working the 09.15 Liverpool Lime Street–Cardiff on 6 October 1989. *Gray Callaway*

diagrams which ran between Paddington and Newbury/Oxford for the day. A pal and I extended our travels down to Slough that day to enjoy further 50s, with a solid day comprising of 47438 from Manchester to Birmingham, 47572 to Oxford, 50025 to Reading, 50032 to Slough, 50031 back to Reading, 50040 to Oxford, 47647 to Birmingham, 85011 to Wolverhampton and 86103 back to Manchester. The short journey from Reading to Oxford provided me with my sixteenth Class 50 for haulage and my only run behind 50040 before it was withdrawn in 1990.

On 16 February the day didn't start well. After a first generation DMU took me from my home station, Sale, non-stop to Manchester Oxford Road on a service from Chester, I immediately returned to Sale and then repeated the five-mile DMU move. I have no recollection of why and assume I returned home because I had forgotten my rover ticket. It was perhaps a blessing in disguise, because when back in Manchester I was very pleased to find 37407 on the 10.00 to Cardiff Central, which I may have missed had I headed south earlier. The tractor was one of the first Scottish 37/4s to be transferred south following

the introduction of Class 156 Sprinters on the West Highland Line. Its transfer from Eastfield to Cardiff Canton depot had only taken place a few weeks earlier and this was followed by 37408, 37411, 37412, 37414, 37422 and 37425, all of which made similar journeys south during 1989. 37407 wasn't in a hurry to lose its Scottish identity and continued to proudly display its "Loch Long" nameplates and large West Highland White Terrier transfer. After a quick move to Stockport to fit in the required 85008, I took the scenic route to the Midlands courtesy of 37407 to Shrewsbury and then made my way to Birmingham for the daily appointment with 50025. That day was the last that I travelled on the Heart of England Rover as for some reason, perhaps illness, the last three of the seven days were not used.

Over the next three months more application was made to my studies in the run up to the GCSE exams in May. Recreational time of course remained important and provided a welcome change to the text books and revision notes, in the form of a handful of days out in the North and Midlands, plus a couple of long-distance trips further south. On the first Saturdays of April and May 1989, I headed to London in search of 50s on the South Western Main Line and found a mixed picture. On April Fool's Day, the unique 87101 took me to Euston and at Waterloo 33109 was on the 10.10 to Salisbury, which I learned was a Class 33 diagram. I have always been very happy to accept new diesel haulages as this was; however, it was Class 50s that I was pursuing. I took the BRCW Type 3 to Basingstoke for 50018 back to Woking (08.17 Exeter–Waterloo) where 47587 took me west (11.10 Waterloo–Exeter).

The day continued well with 50044 (09.40 Exeter–Waterloo), 50003 (13.10 Waterloo–Exeter) and finished with 50017 from Andover to Waterloo (12.20 from Exeter). Instead of taking the usual route from Woking to Waterloo via Surbiton, 50017 took a left turn and proceeded via Staines, Hounslow, Brentford & Putney – the only time I have had a loco-hauled train along that route. Some higher numbered Class 90s had entered service and were starting to appear on passenger trains by then. As the cheaper return tickets from Manchester to London didn't require the holder to specify a particular return service in those days, I would often pick my train based on which had a required loco in charge and break my journey as necessary. On that day I did this by way of 90021 from London Euston to Stafford and 90026 forward to Manchester to increase my tally for the class by two.

A few weeks later, on Saturday 6 May, I had a similar day of Table 145 action (the section of British Rail Passenger Timetable covering Waterloo–Exeter), starting and ending this time in Blackpool. After making my way from Lancashire to London, two return trips to Andover in strong May sunshine saw the Network SouthEast (NSE) liveried quartet of 50002, 50003, 50048 and 50049 in charge of my trains, with "Superb", "Dauntless" and "Defiance" all being winning haulages. The NSE livery had been around for approaching three years and all the 50s except 50002 were in the revised version of the livery by then. 50049 looked smart that day after having recently been renumbered from 50149 following the end of its two-year experiment with lower-geared bogies for freight workings.

Two notable changes took place during mid-May 1989. Firstly I turned 16, which I had mixed feelings

Above: 50018 illustrates the streamlined revised Network SouthEast livery and a hastily attached jumper cable at Woking, while working the 08.17 Exeter–Waterloo on 1 April 1989.

Above: In between some longer-distance travel, after a 12 mile hop from Manchester Piccadilly on 11 March 1989, I caught 37426 in the morning sun as it left Wilmslow on the 10.00 Manchester–Cardiff.

about, as that marked the end of half-price child tickets, although the price increase was mitigated by the benefit of one-third off adult ticket prices with a Young Person's Railcard. On the other hand, however, I was no longer a child and my parents were now happy for me to travel further, including overnight stays, which opened up new possibilities. In addition, this coincided with the commencement of the summer timetable period, which brought the return of Summer Saturdays and holiday trains to the coast. After so many "big engines" had turned out on these services the previous summer, I was excited about their return and what I might find on these. I had bought a copy of the new complete BR Passenger Timetable, so I could see the times of the Saturday trains to and from Skegness and Great Yarmouth for example, but I had no information on the traction that was booked to work these. I therefore planned what I thought would be the best bet for no-heat traction, with a trip to the East Midlands to see what was out on the first Summer Saturday, 20 May 1989.

I made my way to Alfreton & Mansfield Parkway, as the station was known then, by way of a Class

156 from Manchester, where I could then pick up the 10.41 Skegness–Sheffield, as this was the corresponding working to that which produced a pair of Class 20s almost every Saturday the previous year. I was pretty disappointed to find the very everyday 47413 in charge of this, an Immingham-based loco that I'd already had a number of runs behind. The rest of the day's itinerary wrote itself, as a result of the events which then followed. Just after leaving Chesterfield behind 47413, the train was held longer than usual at a red signal and I watched the driver leave the cab and enter a discussion with the signaller. Soon after this we crept forward slowly until reaching the stationary train in front, from which plenty of little heads were also protruding, to try and work out what was happening. It became clear that this was the 09.17 Paddington–York, which had all been a bit too much for 31464 and the 31 had failed. 47413 was coupled to the rear of the train in front and then proceeded to Sheffield sandwiched between the two sets of coaches, propelling one and hauling the other. These were then separated in South Yorkshire and 31107 appeared from nowhere, clearly having been

scrambled as a replacement for the expired 31464. The train was for York only, so not wanting to risk the dreaded penalty fare from the guard, I raced to the ticket office and validated myself with a return to York in time for the no-heat engine north, which was the only interesting working I encountered that day.

Seven days later and still with very little meaningful information on what to expect at the front of the various loco-hauled trains, I had another stab at tracking down new and interesting haulages in the East Midlands. I made my way to Sheffield, to pick up the 08.28 Leeds–Yarmouth this time, which arrived with 47627 in charge. This service had been worked by 47490 the previous week, so the picture that was emerging was not a promising one for finding uncommon traction on passenger trains. I made my way to Derby as I had got wind of some Crewe–Derby trains being loco-hauled, vice the usual DMU. These provided a pleasant trip from Derby to Stoke and back involving 31423 and 31419, followed by a further three Class 47/4s on trains between Derby and Sheffield. My notes from the day tell me that 47215 was hauling a DMU on a Birmingham–Skegness

train and 47331 was on the 12.52 Leeds–Skegness, workings that I most likely learned of when it was too late to reach them. Ironically, when not trying as hard to find interesting haulages the following day, I was more successful when I added two winning NB 47s to my tally, in the form of 47290 (10.19 Carlisle–Leeds) and BR Blue 47284 (14.46 Leeds–Carlisle) while pottering about in West Yorkshire.

3 June 1989 was the third Summer Saturday and "Take Three" of my summer bashing caper. It was another varied day which again I started optimistically without any workings information, with a trip to Derby. I learned that 31223 was on the 10.20 Skegness–Leeds and whilst this was a loco I'd had a couple of times before, I worked out that I could make it to Gainsborough Lea Road in time for a decent 51-mile run behind it, which included some new track for me. Had I had access to someone in the know or a TOPS report for that day, I would have learned that required locos 31282, 31299, 47193, 47298, 47347 and 47372 were all working passenger trains that were accessible to me. The knowledge that I missed

Above: I loved the quiet sleepy atmosphere of Andover station and how the English Electric engines could be heard for miles as they powered into the Hampshire stillness. After taking 50002 from the contrasting bustle of London on 6 May 1989, it is seen leaving on the 13.10 Waterloo–Exeter.

Above: The driver inside 56011 on this southbound MGR working must have wondered what was going on immediately north of Chesterfield on 20 May 1989. To the left 47413 is sandwiched between the coaching stock of both the 10.41 Skegness–Sheffield and the 09.17 Paddington–York, before powering both forward to Sheffield.

these, all of which were within reach that day (as were countless other required locos on other days) doesn't trouble me, as spontaneity and the uncertainty of what one will find on trains were a big part of the hobby's fun and excitement – an aspect that has all but disappeared in recent years.

After limited success at finding new and interesting haulages from the classes that I was fond of, and perhaps because I was sitting my GCSEs, for the next three Saturdays I stayed relatively local and made use of the new North-West Day Ranger. BR had recently introduced this ticket, which for £2.50 provided unlimited travel around the North-West, including the TransPennine services that had recently transferred from Manchester Victoria to Manchester Piccadilly. There was also plenty of interest on the Victoria–Blackpool trains and an

enjoyable Saturday afternoon trip to Liverpool, providing the basher with four Class 31s. At that point I only needed one 31/4 and the trips to Liverpool on those three Saturdays paid off. As Table 26 shows, 11 different 31/4s produced and 31467 provided me with my 68th and final haulage, clearing the sub-class (as 31401 and 31436 had been withdrawn by then and were therefore impossible to reach). When reviewing the 31/4 haulages I had up until that point (May 1986 to June 1989), it was interesting to note that it took 342 individual moves before clearing the sub-class. Whilst I had 31/4s as far as Exeter, Peterborough and the North Wales Coast, I'd travelled behind all but one of the 68 locos between Manchester and Sheffield, 31431 being the only one I didn't have on that stretch, but between Bolton and Blackpool North instead.

Table 26: Double-headed Class 31s passing through Manchester on Saturdays during June 1989.

Date	Locos had on 08.37 Yarmouth–Liverpool	Locos on return working, 16.00 Liverpool–Norwich
10 June 1989	31406 + 31446	31400 + 31411
17 June 1989	31448 + 31425	31438 + 31450
24 June 1989	31408 + 31450	31467 + 31418

Above: The search for diesel haulage on 27 May 1989 led me to the Stoke–Derby route, as "proper trains" were operating vice Sprinters. 31423 leaves Stoke-on-Trent in a hurry on the 11.08 Derby–Crewe, passing at least one pair of Class 20s, several of which were often found stabled in the former bay platform.

CHAPTER 11 –
HEADING SOUTH-WEST

I didn't enjoy it and wasn't greatly motivated at school. Sometimes I simply did the minimum that was sufficient to get by and at other times I applied myself and did my best, particularly when aspects of the subject interested me, or when I could relate to, or was inspired by the teacher. One such man was Mr Mellor, who taught me GCSE English. He was exceptional with teenage boys, knowing how to build respect, rapport, make us laugh and make a point powerfully. He also happened to be a rail enthusiast and had posters of locomotives on his office wall, which the cranks among the pupils would be welcome to peruse. I would doodle in the back of exercise books during lessons, creating front views of diesel locomotives and fictional carriage destination labels for lengthy routes. During one English lesson, instead of engaging with the subject matter I had calculated that during that particular year, I had travelled by rail for an average speed of almost two miles per hour.

The figure was for every hour of the day and night – an average that required something in the region of 15,000 miles per year. He looked over my shoulder during the lesson and instead of reprimanding me for the irrelevant distraction in which I had absorbed myself; "Really, is that true?" he asked, surprised and impressed in equal measure. He then joined in with my deviation from the matter in hand and whilst this was at the expense of the other boys for a couple of minutes, his interest was affirming and left an impression that, with hindsight, I can now see the benefit of.

With the option of leaving school being a genuine possibility at the age of 16, I had applied for a number of jobs during the first half of 1989. At the time many firms were employing school-leavers on one of the many Government-backed Youth Training Schemes (YTS), which provided on-the-job training for 16 to 18 year olds. The incentive for the teenager was the

Above: On two out of three consecutive days 31178 provided the entertainment on the 08.15 Nottingham–Blackpool. With Warbreck Water Tower just visible at the far left, the train relieves itself of more holiday-makers than loco bashers at its destination on 20 June 1989.

Above: When this photograph was taken, only a few weeks remained before Class 33 diagrams on the stopping services from Waterloo would be surrendered to Class 50s. Most services terminated at Salisbury and often consisted of hauled multiple units rather than coaching stock, as seen above when 33102 brought the 12.10 Waterloo–Gillingham into Woking on 1 April 1989.

opportunity to start what could be a long career in one's chosen industry and for the employer, it may have been the very nominal wage of only £27.50 per week – at a time when the average earnings were over £200.00 per week! One YTS I applied for was with British Rail. I was successful and could have begun a career on the railway; however, I was distinctly unimpressed with how run-down the interior of Rail House on Manchester's Store Street was and what I perceived to be a tardy culture among the staff I encountered during the recruitment process. Had I made a decision with my heart instead of my head, the following decades would no doubt have been very different. I expect I would have worked towards becoming a driver and most likely be working in the rail industry today. Another YTS that I applied for was with a large engineering firm and I recall not being inspired by the many machines I saw in the workshop as much as would be expected for a young man who loved diesel engines! During the interview, my answer to a routine question about my interests had probably been more detailed than necessary, so much so that the interviewer's face suddenly lit up and he interrupted me and said "I know what you

are – you're a basher!" It transpired that he knew another basher and the interviewer then proceeded to relay how this person would travel all over Britain chasing diesel engines. He reeled off various locations that this chap had been to and I recall this included Fort William, which of course was a well-known destination for Class 37 fans in the late 1980s. The two other people assisting him with the interview looked on completely bemused by the unexpected turn the proceedings had taken!

Despite the appeal of leaving school at 16 and having been for a number of interviews, some of which had yielded job offers, by the time I had sat the last of my GCSE exams around mid-June, I reluctantly conceded that the option with the best long-term prognosis was to remain at Sale Boys' Grammar School. I made my intention to continue studying Geography, Biology and Chemistry to A-Level known to the school and then began a very long school summer holiday, with the contentment that comes from knowing that no lessons or studies will feature for three months.

Bashing-wise, I found 31178 on the 08.15 Nottingham–Blackpool on both Tuesday 20 and

Thursday 22 June and so I took the Goyle for 55 miles from Stockport to Blackpool on both dates. I always loved the atmosphere of this route, particularly between Preston and Blackpool, complete with signal boxes and semaphore signals which managed to last almost another 30 years. The summer's big adventure however began the following week, when my bashing mate from Sheffield joined me in heading south, both of us equipped with a massive rucksack containing supplies to last us eight days. Being fairly organised and wanting to make the most of this new level of independence, much planning had been put into this trip. Apart from a Freedom of Scotland, the only rail rover I hadn't done was a South-West, which was where our beloved Class 50s and a host of other diesel haulage could be found. We had written to some tourist information offices some weeks earlier, asking that they send us details of the less expensive accommodation in the region, as this was the only way to access such information in the pre-internet days. In the end our budgets prescribed that we were to stay in Youth Hostels, positioning ourselves in Salisbury at the eastern end and Penzance at the western end of the rover's area of validity.

We knew that the beginning of the end had begun for Class 50s, with a fifth of the class already having been withdrawn and I'd only had 23 of the 50 that were built at that point. The early withdrawals meant that 50006, 50010, 50011, 50013, 50014, 50022 and 50047 were beyond reach and all of those except 50010 and 50011 had been scrapped by July 1989. Class 50 workings to Newquay were long gone and 50 haulage to Penzance was fairly uncommon by then. One positive for the class was that from the recent May timetable change, 50s had taken over the Waterloo–Salisbury stopping services, which were previously in the hands of Class 33s. These additional workings required more Type 4s to be in service and so a number of 50s had been transferred from the DCWA pool to the NSSA pool, which supplied locos for the Table 145 (Waterloo-Exeter) trains. The DCWA pool was generally populated by 50s with a higher number of engine hours, which were restricted to a maximum speed of 60mph and less demanding engineering duties. In haulage terms, the DCWA 50s were comparable to NB locos and were very rarely used for passenger trains. One of the engines making the welcome return to the NSSA pool was 50016, which we both needed for haulage, so this was high on our "To Do" list for the week!

We decided to start by tagging an extra day on to our travels to fit in some 50s in the South-East,

Above: The driver of 50005 on the 17.35 Paignton–Waterloo keeps a close eye on proceedings while giving me enough time to jump out for a quick photo at Dawlish before continuing to Salisbury on 2 July 1989. Note how close the red safety line is to the platform edge, a distance which has increased in the years since this photograph was taken.

so after heading to London in the morning, we got down to business with 50035 from Paddington to Oxford and back (12.15 Paddington–Oxford & 14.00 return), followed by 50026 back to Reading (15.15 Paddington–Oxford). I loved the sense of space on the long rakes of Mark 1 coaches, which were sparsely populated during the daytime, and the speed with which the 50s would race through the urban sprawl. We then took another fine sounding English Electric engine, albeit a smaller one, in the form of a Class 205 diesel-electric multiple unit from Reading to Basingstoke. From there I made the most of the outward leg of my Manchester–Salisbury ticket, courtesy of the evening stopping trains. These involved 50041, 50029, 50017, 50003 and 50050 for short runs west, the likes of which included stops in Overton, Grateley, Andover and Salisbury. That first day was highly productive, with five of the seven Class 50s being new to me, taking my haulage total for the class over the half way mark, reaching 28.

Salisbury Youth Hostel was our base for the first night and we found beds in the large male dormitory, which was housing travellers from all over the world.

We learned that many of these had reached Salisbury specifically to see Stonehenge, something that hadn't even occurred to us, with an attraction of an entirely different kind occupying our minds. It was a hot summer night and I vividly remember lying in the bunk beds after "lights out" with the dormitory window wide open, hearing a beautiful sound while all was quiet. The Youth Hostel sat on a hill with the railway about a mile away in the valley below and the hammering of an English Electric 16CVST resounding in the summer midnight air was a joy to behold, as I drifted off to sleep.

The next morning was Friday 30 June and Day 1 of the South-West Rover. This started with 50041 from Salisbury to Exeter (06.52 Waterloo–Exeter), followed by an HST forward to Plymouth and a Class 101 DMU along the Gunnislake branch, as exploring new lines has always featured in my pursuits. When 50002 arrived from Laira with the 14.17 Plymouth–Portsmouth, there was only one thing to do; however, at Exeter we then needed to head west again, as we had booked to stay in Penzance Youth Hostel that night, hoping to position ourselves in the path of a 50 out of

Above: In recognition of its historical significance, the first Class 37 to be built had its D6700 number and 1960 BR green livery reapplied in 1988. 37350 takes a breather at Weymouth after working the 16.54 from Bristol Temple Meads on 3 July 1989.

Above: Tisbury in Wiltshire is one of many locations that I would not have visited or passed through were it not for the diesel engines found there. On 4 July 1989, 50009 sets off towards the passing loop at the east end of the station, at the head of the 17.37 Exeter–Waterloo during the low evening sun.

Penzance the next morning. Along with the 50s we required, a member of the class to or from Penzance was at the top of our aims for the week, as we felt that it was "now or never" for a chance of doing so. We learned that our best bet was the Saturdays-only 10.20 Paddington–Penzance and return Sunday working, the 10.35 Penzance–Paddington, as this had yielded a 50 on some Saturdays and a 47 on others. So on Saturday 1 July, when there was no suitable loco-hauled train leaving Penzance, we took an HST for the 131 miles to Exeter, which at that time was still single track in places. As we passed through Dawlish Warren, we overtook ex-DCWA pool 50016, which was on our hit list and in charge of the 09.33 Plymouth–Brighton. We duly alighted the HST at Exeter and waited for "Barham" to arrive and run round, before bashing it for the 54 miles to Sherborne. That turned out to be my only run behind 50016, which was withdrawn the following year. After 50029 took us back to Exeter on the 09.05 Brighton–Plymouth, we were comfortably ahead of the 10.20 Paddington–Penzance and desperately hoping this would produce a 50. While

50029 was running round the stock at Exeter St. Davids, another interesting distraction appeared, in the form of 31405 and 31454 on the 09.33 Stockport–Paignton. Having always had a soft spot for these machines, we squeezed in a short run behind them as far as Dawlish. We then resumed our journey west on the same Brighton–Plymouth train behind 50029 as far as Newton Abbot, to await the anticipated Penzance train. The day continued to go as well as we hoped when large logo liveried 50036 rounded the curve and paused under Newton Abbot's ornate GWR canopy. We thoroughly enjoyed the run to Penzance and were up and down from our seats to savour the thrash from the front window while departing from some of the many stations at which we stopped through Cornwall.

After arriving into Penzance behind 50036, which I last had at the other end of the country on a run from Glasgow the previous year, the evening was spent exploring the St Ives and Falmouth branches and their picturesque resorts. Back at Truro, we picked up 47459 on the "Up Midnight" sleeper and embarked upon our first overnight. Having a ticket that is valid through the night provides the opportunity to both save some money on accommodation by sleeping on trains and increase your diesel mileage. That's the theory, but in reality I have always struggled to sleep on trains and ended up worse for wear the following day, so I've done relatively few overnights through the years. We had a long wait at Bristol Temple Meads in the early hours and because it was pretty cold during those small hours, we bedded down in the station's main subway using our rucksacks as pillows. The few railway staff that passed us didn't seem to mind; I can't imagine that would be the case today! We then took 47459 for the 207 miles from Bristol to Penzance and did our best to get some sleep along the way. That Sunday turned out to be another excellent day, with 50036 from Penzance to Exeter (10.35 Penzance–Paddington), for an HST to Totnes and then 50030 back to Exeter (14.42 Plymouth–Waterloo). After making our way to Paignton on that warm, sleepy afternoon, we were then in position for our third Class 50 of the day through the beautiful scenery

between Newton Abbot and Exeter. The Teignmouth–Dawlish section includes the sea wall walkway, which is sandwiched between various beaches and the railway. On a Sunday in July, hundreds, if not thousands of people can be seen walking along this, intrigued by the many heads that would protrude from the windows of the passing diesel-hauled trains. Sometimes photographers could be spotted at the lineside or in elevated locations such as on Langstone Rock and some would give them a good bellow to endorse the traction. With the elimination of droplight windows and an increased attention to safety following some tragic incidents in recent times, those antics are now consigned to history. I needed 50005, which took us on a wonderful 117 mile run from Paignton to Salisbury. The front coach was loaded with enthusiasts and after being blasted by the sea air through Dawlish and minimal sleep the previous night, I lay across two seats and drifted in and out of sleep, to the soundtrack of Class 50 thrash through Devon, Dorset and Wiltshire.

Above: On the afternoon of our final day in the South-West, DCWA Pool 50021 trundled into Exeter St. Davids on an engineers' working from the west. After depositing its wagons, it tantalised us further by entering the eastern bay platform, but sadly didn't get any nearer than this to the 14.22 to Waterloo on 6 July 1989. The loco is still required three decades later; however, thanks to a huge amount of recent restoration work which is now approaching completion, an opportunity to be properly acquainted with "Rodney" should arise soon.

After a better night's sleep in Salisbury Youth Hostel, a large part of the Monday was spent travelling behind 50s between Salisbury and Exeter, with locos including 50041, 50009, 50017, 50001, 50027 and 47557. Of these, 50001 and 50027 were the last of the winning haulages for the week, taking my total for the class to 32 and mileage for the class also received a decent boost. We then took an HST from Exeter to Bristol, which included the required track via Weston-super-Mare. In Bristol, we stocked up on nutritious supplies, which included a massive three litre bottle of beer and packets of custard cream biscuits – I know this as I have photos of us drinking the beer and I remember an argument, which led to a packet of custard creams being thrown and hitting my head! The reason for travelling to Bristol was to do the Class 37 hauled trains between Bristol and Weymouth and on 3 July 1989, 37350 was in charge of the 16.54 Bristol–Weymouth. This is an historically important loco, being the first Class 37 to be built and it can now be seen in the National Railway Museum, York as part of the National Collection. It began its life as D6700 in late 1960, was renumbered to 37119 under TOPS in 1974 and then to 37350 in March 1988 when it received re-geared bogies. A few months later it was rebuilt and its 1960 green livery was reapplied, complete with its 1957 series D6700 number and that's how we found it. It had been in charge of many passenger trains to and from Bristol since May that year and provided us with a very pleasant trip to Weymouth and then back to Yeovil Pen Mill (19.44 Weymouth–Bristol). Neither of us had visited Yeovil before and had allowed around 20 minutes to walk between Pen Mill and Junction stations. We'd had a conversation that went something along the lines of "Yeovil isn't a very big place, so the two stations can't be that far apart and 20 minutes to walk between the two should be enough". After 37350 powered away from the photogenic Pen Mill station, we followed the signs to Junction station and soon found ourselves walking along a country road in the middle of nowhere after 9.00pm. As the day turned into dusk, the familiar sound of a 50 accelerating on the last Exeter–Waterloo service of the day was, for once, not music to our ears. We learned two useful lessons that day; more than two miles separate the railway stations in Yeovil and sufficiently detailed advance planning is always worthwhile. Fortunately there was a final eastbound train to take us to our beds in Salisbury, in the form of the 21.30 Yeovil Junction–Salisbury, which was in the hands of 50050. We were surprised to find that this was effectively our own private charter, as we walked the length of the seven-coach train and discovered the guard was the only other person on board. He didn't seem to leave his brake compartment for the entire duration of the journey!

For the last three days of the week-long rover, one was an enforced rest day, as no trains operated on Wednesday 5 July 1989. This was one of a series of weekly dates when BR staff were on strike over the renumeration increase they had been offered. The matter was settled a few weeks later, when an 8.8% pay rise was accepted. That may sound excessive in the context of the financial climate three decades later, however that was at the time when the Bank of England base rate was a massive 14.875%, as opposed to being below 1%, as it has been for more than a decade at the time of writing in 2019. We therefore spent the enforced break doing what most normal people do when they travel – exploring Salisbury, walking along the River Avon and taking in views of the country's tallest church and cathedral spire. It seems that I didn't consider anything non-railway related worthy of recording by way of a photograph, though!

With the exception of a run from Exeter Central to Exmouth and back on a Class 101 DMU, the last two days were spent accruing Class 50 mileage between Salisbury and Exeter, with changes at various examples of the rural stations along the route. One such stop was at Templecombe, where after photographing 50003, watching and listening as it broke the Somerset silence along the single track towards Exeter, I panicked. I realised I'd left my rucksack on the train and this contained everything needed for the week and most importantly, my moves book and Motive Power Pocket Book, which between them detailed all my records. Templecombe station had reopened a few years earlier and was unstaffed; however, I had an idea. At the "down" end of the platform was a signal and from the times I had observed drivers leave loco cabs when held at red signals and speak with signallers, this could offer a way to mitigate my potential loss. With no idea how to operate a telephone that didn't have any buttons or controls and I picked up the handset and said "Hello?" When a voice responded, I explained that I was a passenger on Templecombe station and the predicament I found myself in. To my relief, the helpful voice assured me that he would arrange for someone at Exeter to retrieve my effects, which I was gratefully reunited with later that day.

We were very content after seven days of diesel bashing, which included 18 different Class 50s

(50001, 002, 003, 005, 007, 009, 016, 017, 026, 027, 029, 030, 035, 036, 041, 048, 049 & 050), members of Classes 31, 37, 43 (HST) and 47. The week was rounded off well by spending over six hours in the company of 50002, which we enjoyed for the 260 miles from Salisbury to Exeter and then from Exeter to Waterloo. Since then, and as I increasingly accumulated mileage behind "Superb" over the next few years, it has remained a firm favourite. I was delighted when it was preserved; the loco is now undergoing a long-term overhaul at the South Devon Railway where it is based. After 87034 returned me to the North, it was a whole nine days before I was out again seeking haulages.

CHAPTER 12 –
CHANGING TIMES

After returning from exploring the South-West at the start of July, two months of the long summer break still remained before I was due to begin the sixth form. Perhaps because I had a bit too much time on my hands and limited remaining funds for further travel, or maybe it was just the tumultuous teenage years, but at the time there were some tensions at home between myself and my parents. I loved staying with my grandparents in Blackpool, so I headed up there for an extended summer stay from mid-July. A change is as good as a rest and this also opened up a couple of new opportunities.

The 08.15 Derby–Llandudno was a train that saw an interesting variety of traction through the summer of 1989. On weekdays, after reaching Llandudno the loco would make the scenic journey to Blaenau Ffestiniog (departing Llandudno at 12.04 and returning from Blaenau at 14.18), before travelling back to Derby on the 17.27 from Llandudno. On Saturdays, after running round at Llandudno, it would return earlier, on the 12.47 Llandudno–Derby. An ETH-fitted Class 31 was the most common motive power; however, this train also enjoyed an eclectic mix of engines that year, including 97204 (formerly 31326) on 24 June, 97480 (formerly 47480) on 26 June, 47215 on 11 July, 47187 on 13 July, 31312 on 19 July, 97545 (formerly 47545) on 14 August, 31206 on various dates in August and 31232 on 16 September. On occasions it produced a pair of 20s, such as 26 August, when it started off behind 20127 & 20114, which were replaced at Crewe by 20071 & 20182 after 20127 failed. Another such Saturday was 15 July, when I happened to make a journey to Crewe in order to sample this train. I was surprised to find the coaching stock comprised of a 3-car Class 117 DMU, but the healthy number of bashers on board didn't

Above: Class 20s were one of a number of locomotive types used on the daily Derby–Llandudno service during the summer of 1989. On 15 July 20005 & 20208 are at seen at the North Wales terminus after having arrived hauling a 3-car Class 117 DMU on the 08.15 from Derby.

Above: On a day when all kinds of traction was being thrown out on relief trains to the North as a result of problems with the overhead wires near London Euston, 20108 & 20215 reach the blocks at Manchester Piccadilly with the 10.00 from St Pancras on 21 July 1989.

mind, as there were still plenty of internal opening windows to provide us with ventilation and stereo thrash. The locos were the required 20208 partnered with 20005, which I'd had a couple of times on Skegness trains with 20048 the previous summer. I took these along the picturesque North Wales Coast Line to Llandudno, which still had a comprehensive station roof at the time, and then back to Crewe. I also managed to add haulages from 85018, 87011, 87012, 31439 and 31467 to the activities before calling it a day.

After a day of rest on the Sunday, I made my way up to Blackpool on the Monday to begin a stay of one and a half months. It soon became obvious that having my bike would be very useful, for getting around generally and also because I immediately landed a paper round delivering the morning daily titles. A few days later therefore, on Friday 21 July, I made a trip back to Sale to collect my bike. As there was a practical purpose to my train journey for a change, rather than diesel bashing, I wasn't expecting much in the way of unusual workings. Out of all the years I have been pursuing diesel locos, that day

turned out to be my luckiest and the best example of being in the right place at the right time for falling upon rare haulages. The day started well with the required no-heat 31206 from Layton on the 08.35 Blackpool–Stockport stopping service. At Preston, fellow bashing associate Andrew from Bamber Bridge boarded and joined me in the Mark 1 corridor coach's compartment in which I was relaxing. After then disappearing for a few minutes, he returned and with a cheeky grin and asked whether I needed many 56s? At that point I'd never had a 56 and I suspected he hadn't either, so we both knew that this was a rhetorical question. He had gleaned some intriguing information from other bashers on the train, learning of a problem with the overhead wires near London Euston, which had resulted in there currently being no trains in or out of the terminus. Consequently, BR had organised some relief services to the North-West, departing from nearby St Pancras, running via the Midland Main Line and across the Hope Valley route. These were utilising whatever traction was available and we understood this included 56004, which we could reach if we continued behind

31206 to Stockport. There was also a pair of 20s on another relief service that was also on its way to Manchester. The day's agenda had suddenly become more complicated!

On arrival at Stockport, we found around a dozen of the regular bashers who had somehow also heard the news and were poised for some required haulage. One or two of these stood out because instead of being in the usual low-key attire, they were dressed in suits. These were the fortunate "top-link" bashers who had informant friends in high places and were also able to duck out of their workplaces when a particularly "large" working was taking place! It wasn't long before the rumour became reality and the fourth Romanian Grid to be built became my first Type 5 haulage, working the 07.10 St Pancras–Manchester Piccadilly relief. In between the giddiness arising from the day's drama, I needed to return to Sale and fulfil the original purpose for my travel. After doing this and unwisely not collecting my bike lock, I was then lumbered with a large racer, which I could have otherwise locked to some railings in Manchester and travelled to Stockport much lighter. I returned to Stockport as soon as possible in pursuit of the rumoured Class 20s and it wasn't long before my bike and I were on another relief working between Stockport and Manchester. This time Toton's red stripe Railfreight liveried 20215 and 20108 were in charge, providing me with another two required engines. After the drama of 31206, 56004 and a pair of 20s into Manchester Piccadilly, it was then straight back down to earth with a bouncy ride on a Class 142 all-stations stopping service to Blackpool, doing my best to stop the bike from falling over!

My grandparents lived at the highest point in Blackpool, immediately opposite the second most well-known of the town's towers – Warbreck Water Tower. I often did a double paper round, which involved some demanding ascents and descents up and down the steep hill leading to the tower, something I wasn't used to with being from an area with completely flat topography. Trains, and more specifically diesel haulages, were never far from my mind and I soon developed a productive morning routine. This consisted of beginning the paper round at about 07.00, so I could finish it in time to be at Layton station to see what was working the 06.55 Manchester Victoria–Blackpool. I kept my records book with me and if it was something I needed, I would then peddle fast for the 1.25 miles to North Station, buy a single to Layton and just about have time to load my bike onto a carriage before the

returning loco left on the 08.35 Blackpool–Stockport. Alternatively, if I didn't fancy the cycle sprint, I would wait for the loco to return to Layton around 20 minutes after it first passed through, and do it for the equally, "ned", two miles to Poulton-le-Fylde. It was remarkable how often the traction on this train changed and as Table 27 shows, these exploits yielded four new 31s in the form of 31292, 31124, 31123 and 31286 on the dates shown. Once back at Layton, after a final ascent of the steeply-graded Warbreck Hill Road, I was then more than ready for the large cooked breakfast that awaited me.

Table 27: Locos that worked the 06.55 Manchester Victoria–Blackpool North and 08.35 Blackpool–Stockport during the summer of 1989.

Date	Loco	Date	Loco
Mon 17 July	31237	Mon 14 Aug	31412
Tue 18 July	Unknown	Tue 15 Aug	31402
Wed 19 July	Unknown	Wed 16 Aug	31431
Thu 20 July	Unknown	Thu 17 Aug	31286
Fri 21 July	31206	Fri 18 Aug	31101
Mon 24 July	31232	Mon 21 Aug	31437
Tue 25 July	31232	Tue 22 Aug	DMU
Wed 26 July	Rail Strike	Wed 23 Aug	31431
Thu 27 July	31417	Thu 24 Aug	Unknown
Fri 28 July	31446	Fri 25 Aug	Unknown
Mon 31 July	31416	Mon 28 Aug	DMU
Tue 1 Aug	31292	Tue 29 Aug	DMU
Wed 2 Aug	31409	Wed 30 Aug	31464
Thu 3 Aug	31409	Thu 31 Aug	Unknown
Fri 4 Aug	31409	Fri 1 Sep	Unknown
Mon 7 Aug	31409		
Tue 8 Aug	31124		
Wed 9 Aug	31123		
Thu 10 Aug	31445		
Fri 11 Aug	31101		

One of the news headlines on 7 August 1989 caught my eye and greatly concerned me. The previous evening's 21.15 Oxford–London Paddington had been derailed at West Ealing, with the television and newspaper pictures clearly showing an overturned Network SouthEast liveried

Above: After covering the early morning Manchester–Blackpool working and finding 31101 in charge on 18 August 1989, I was back at Layton station again that evening to find 31292, where it is seen on the 17.10 Manchester Victoria–Blackpool North service.

Class 50, which had suffered significant damage. The leading coach had ridden onto the loco and most of the other coaches were derailed. Miraculously no one was killed in this incident, which it turned out was the result of vandals placing a rail on the track. I poured over the television reports and newspaper pictures, desperately trying to work out which loco was involved, as I feared for its future, but wasn't able to identify it. In time I learned that this was 50025, which was soon moved to the adjacent supermarket car park and subsequently to Old Oak Common depot, where it was scrapped later that year. "Invincible" was a loco I'd had seven runs behind, all on the route between Paddington and Birmingham, the last of which was in February 1989. Sadly it did not live up to its name.

There were a lot of 31 hauled Class Two stopping services in the North-West that summer. Many were in the hands of 31/4s, which had been freed up when their duties on the Liverpool–Sheffield (and beyond) and Birmingham–Norwich services were taken over by Sprinters the previous year. In the warmer months, though, no-heat 31/1s were fairly common, plying their trade on diagrams which included Manchester to Blackpool and Barrow, as well as the Liverpool to Preston and Manchester stopping trains. I'd cleared the 31/4s by then, but only had around 30 of the more than 150 Class 31/1s that remained in service at that point. I've always enjoyed 31 haulage and whilst it wasn't entirely about scoring new engines, this was still a big attraction and there were still plenty of locos to go at. As Tables 28 & 29 show, two North-West

Below: The day before the disaster which put it on the front pages of the national newspapers, 50025 worked the 13.02 Paddington–Paignton and 17.45 Paignton–Paddington. Oblivious to what awaits the following day, on 5 August 1989 50025 calmly passes Great Cheverell on the outward working to Paignton. *Douglas Johnson*

Day Rangers on consecutive Wednesdays in August and a fortunate find on 22 August provided plenty of entertainment in this area and where the workings are known, these have been added – note the two no-heat 31s on Class One trains in the second table.

Another train that caught my attention while in Blackpool was a semi-regular relief which ran on weekdays from Blythe Bridge to Blackpool and was hauled by two Class 20s. It would depart Blythe Bridge at 09.35 and travel via Stoke-on-Trent and Crewe. The return working left Blackpool at 18.42 and as its first stop was at Crewe, this was a difficult working to do, because it would mean not being back in Blackpool until after 22.00. I watched this depart from Blackpool on a few occasions, ruefully letting some required engines go without me on board and these included 20078 & 20040 (29 August) and 20023 & 20141 (30 August), none of which I had another opportunity to have before they were scrapped. There was a second semi-regular Class 20 hauled relief from Blythe Bridge, which often ran on the same day as the Blackpool train, consisting of the 08.15 Blythe Bridge–Llandudno and 18.10 Llandudno–Blythe Bridge. I didn't encounter this service, but it also produced a very good variety of the class, which included 20090 & 20120, **20099 & 20218**, **20141** & 20169, 20032 & 20063, 20005 & 20208, 20071 & **20182**, 20189 & 20056 and **20139 & 20160** – the locos in bold being the examples I have had for haulage.

On the last day of August, I returned home by way of 31431 from Layton to Deansgate, where I walked through the ornate Victorian subway and made my way to Sale on a Class 304. In a number of ways, this journey marked a time of change; it was the end of the long summer of 1989, my extended stay in Blackpool and the last time I travelled by train for a number of weeks – an almost geological period by my standards. I was by no means finished with loco bashing, but a number of new responsibilities and activities were just around the corner and these would begin to encroach on what until that point had been

Table 28: A productive day dominated by Class 31 haulage.

Date	Traction	From	To	Mileage	Details where known
16-Aug-89	31431	Layton	Preston	16.25	08.35 Blackpool–Stockport
16-Aug-89	86427	Preston	Lancaster	21.00	WCML service
16-Aug-89	31464	Lancaster	Manchester Victoria	51.75	Barrow– Manchester service
16-Aug-89	31464	Manchester Victoria	Eccles	4.00	?? Manchester–Liverpool
16-Aug-89	31101	Eccles	Manchester Victoria	4.00	11.27 Liverpool–Man Victoria
16-Aug-89	31101	Manchester Victoria	Liverpool Lime St	30.75	12.59 Man Victoria–Liverpool
16-Aug-89	142	Liverpool Lime Street	Wigan North Western	20.00	
16-Aug-89	31272	Wigan North Western	Broad Green	16.50	15.47 Preston–Liverpool
16-Aug-89	150	Broad Green	Manchester Victoria	28.25	
16-Aug-89	150	Manchester Victoria	Wigan Wallgate	18.00	
16-Aug-89	87020	Wigan North Western	Preston	15.25	WCML service
16-Aug-89	47623	Preston	Poulton Le Fylde	14.25	Euston–Blackpool service
16-Aug-89	150	Poulton Le Fylde	Layton	2.00	

Table 29: Another productive two days dominated by Class 31 haulage.

Date	Traction	From	To	Mileage	Details where known
22 Aug 89	150	Layton	Preston	16.25	
22 Aug 89	31431	Preston	Manchester Victoria	30.75	
22 Aug 89	31249	Manchester Piccadilly	Oxford Road	0.50	12.23 Newcastle–Liverpool, after 47802 failed
23 Aug 89	31431	Deansgate	Stockport	7.00	
23 Aug 89	90009	Stockport	Manchester Piccadilly	6.00	
23 Aug 89	47606	Manchester Piccadilly	Bolton	11.25	
23 Aug 89	31123	Bolton	Manchester Victoria	10.75	09.27 Barrow–Manchester Victoria
23 Aug 89	31123	Manchester Victoria	Eccles	4.00	11.59 Manchester Victoria–Liverpool
23 Aug 89	31101	Eccles	Manchester Victoria	4.00	11.27 Liverpool–Manchester Victoria
23 Aug 89	31101	Manchester Victoria	Huyton	26.25	12.59 Manchester Victoria–Liverpool
23 Aug 89	31433	Huyton	Preston	29.75	Not 100% this was 31433
23 Aug 89	156	Preston	Manchester Piccadilly	31.25	
23 Aug 89	Unknown	Manchester Piccadilly	Stockport	6.00	
23 Aug 89	31224	Stockport	Manchester Piccadilly	6.00	13.34 Reading–Manchester
23 Aug 89	150	Manchester Piccadilly	Layton	47.50	

Left: The Trans-Pennine services between Newcastle and Liverpool were in the hands of Class 47/4s in 1989; however, on 22 August the Type 4 had failed and had to be replaced by whatever could be found in the Huddersfield area. 31249 was summoned and is shown leaving Manchester Oxford Road on the 12.23 Newcastle–Liverpool, which by then was more than an hour late.

my sole interest. A few days later I would return to school and begin studying towards A-Levels, which required me to up my game a little. A week or so later, I landed a regular job at a petrol station in Sale. It was unique in the locality in that the driver didn't leave their seat and I would fill the vehicle up and take the money – with the responsibility of not filling it above their prescribed amount falling on me. I had regular shifts after school and on alternate Saturdays starting at 13.00, boosting my income but limiting availability for travelling.

September 1989 was also the time when two new passions started to take a hold. Rock bands such as Fleetwood Mac, Def Leppard and Iron Maiden had firmly caught my attention and I also acquired my first guitar. I began guitar lessons, increasing in proficiency and started to write the first of what would become many songs, some of which would later find an audience. I also sampled my first live music concerts that autumn, in the form

of Transvision Vamp and Xentrix, which perhaps represent the two ends of the guitar-based music spectrum that I still love. As if all of that wasn't enough to occupy me, that was also the time I first started attending the local Baptist church, where I found some fantastic new friends and clarity to the faith which already existed, but had perhaps become a little dormant.

All of this together left me with less time for bashing. The only trip of note that I made in September is summarised in Table 30. This was a good day for diesel haulage, although the 90 was the only required loco. While returning north on the long journey behind 47426, I was listening to the radio on my Walkman and was delighted when I heard that Manchester City had beaten their local rivals 5–1 in the Derby. This was an exceptional result for City, at a time when the two clubs were poles apart, and marked the beginning of my support for them, which in time would grow.

Above: 31272 was one of many members of the class to work local stopping trains in the North-West on 16 August 1989 – a day when train heating was clearly not required! It is shown leaving Broad Green while working the 15.47 Preston–Liverpool Lime Street.

Above: Aware that unusual traction on service trains would draw plenty of enthusiasts, BR arranged a variety of interesting loco haulage on the Settle & Carlisle line on Saturdays during November 1989. On 25 November, 20061 & 20093 (with 47444 used for train heating) provided me with over 225 miles of Type 1 action. They are seen heading south with the 12.42 Carlisle–Leeds. *Douglas Johnson*

Table 30: Moves on Saturday 23 September 1989:

Loco	From	To	Mileage	Details
90025	Manchester Piccadilly	Birmingham International	90.50	
50033	Birmingham International	Wolverhampton	21.25	07.05 Paddington–Wolverhampton
50033	Wolverhampton	Reading	111.75	11.18 Wolverhampton–Paddington
50026	Reading	Oxford	27.50	15.15 Paddington–Oxford
50035	Oxford	Reading	27.50	17.00 Oxford–Paddington
47436	Reading	Manchester Piccadilly	181.00	15.15 Folkestone–Manchester

November saw a burst of bashing when BR organised a series of workings on consecutive Saturdays, using locos which were rare for haulage on the Settle and Carlisle line. These attracted hordes of enthusiasts and as the traction was all of the no-heat variety, a Class 47 was tucked behind the leading engine(s) to provide train heat during that colder month. The locos worked the 08.25 Leeds–Carlisle & 12.42 Carlisle–Leeds; with an early start from south Manchester, I could make it to Leeds in time to "scratch" the loco from Leeds to Keighley and then return to Sale in time to start work at 13.00. The trains that November provided rare haulages from 56104 (on 4 November with 47503 for train heat), 56030 (11 Nov & 47475 for heat) and 56099 (18 Nov & 47477 for heat). When I learned that it would be a pair of 20s on 25 November, I decided that was worth the full run to Carlisle and back; double winners 20061 and 20093, with 47444 for train heat, fulfilled that duty.

That brought a period of great transition for me and the formative decade of the 1980s to a close. Looking back, it's easier to highlight landmarks and turning points, but at the time I was fully occupied with questions about which bands had an album out next, what the current loco-hauled diagrams were, what I believed about social issues and whether I should go to university. The answers to questions such as those would seep out in different ways over the years to come and in terms of loco bashing, more chapters were yet to be written.

APPENDIX I – GLOSSARY OF BASHING TERMINOLGY

Bashing is a niche interest which has developed its own jargon over the years. Below is a light-hearted look at some of the more commonly used terms and this is by no means an exhaustive list of the terminology. There are plenty of other words and phrases and there will no doubt be more that I haven't come across. The list below doesn't include locomotive class nicknames, many of which are listed in Appendix 2.

Basher	Pursuer and collector of locomotive haulages.	**Minus five**	A connection whereby the next timetabled service you hope to catch departs five (or any other number of) minutes before the incoming train is due to arrive.
Beast	A term of affection for a particular loco.		
Bellowing	Extending of the arm and associated hand waving from the train window, to indicate one's appreciation of the traction. This was common in the 1980s and is no longer acceptable behaviour.	**NB**	No-heat locomotive. Originates from the days of steam heating – "No Boiler".
		Ned	Abbreviation of New Engine Desperado – referring to bashers who tend to prioritise new haulages. It can be a negative term, however bashers all started that way and who doesn't enjoy a winning loco?!
Bert and Ada	Respective generic terms for male and female non-bashing travellers.		
Clag	Exhaust emitted from the diesel engine, providing visual embellishment to the thrash.		
Cleared	Having completed a particular action for all members of a specific class or sub-class (e.g. seen or hauled-by all).	**Normals**	The average Joe Public rail traveller who is not a rail enthusiast.
		Overnight	Continuing to travel by rail through the night, attempting to sleep on trains or stations rather than in a bed, in order to maximise travel.
Crank	The rail enthusiast, generally referring to bashers.		
Dreadful	In opposition to its usual meaning, a complimentary term for a loco or experience.	**Piece**	Rail ticket or rover.
		Plus two	A connection whereby the next timetabled service you aim to catch departs two (or any other number of) minutes after the incoming train is due to arrive.
Dud	A not-required loco.		
ETH	Locomotive fitted with Electric Train Heating.		
Footex	Abbreviation of Football Excursion; a special train to transport football fans to a particular match. Other similar terms include the Rugex (Rugby excursion) and Hippy-ex (special for a music event or festival) etc.	**Roadshow**	A collection of Cranks. Often used to describe a group following a particular class of locomotive.
		Scratch	This means the same as Score.
		Score	The action of having a new haulage.
		Seminar	A gathering of bashers in front of a train loco for the purpose of photographs, something that is rarely seen today. Seminars may or may not involve stationary bellowing actions.
Gen	Useful advance information about which locos are working on particular services.		
Gripper	Ticket inspector (grip being the verb).		
Hellfire	Highly positive adjective for a loco, usually due to the sound it makes.	**Thrash**	The powerful sound of a diesel locomotive working hard – music to the basher's ears.
Moves	A sequence of individual rail journeys.	**Winner**	A required loco..

APPENDIX 2 –
LOCOMOTIVE NICKNAMES

Below are some of the locomotive class or sub-class nicknames which rail enthusiasts can be heard using. Bashing can be a little territorial in some quarters and some of the nicknames are generally considered derogatory – these are underlined.

08	Gronk	50	Hoover, Vac, Dubber Log
09	Super Gronk	52	Western, Whizzo, Thousand
14	Teddy Bear	55	Deltic
17	Clayton	56	Grid
20	Chopper, Bomb	57	Bodysnatcher
23	Baby Deltic	58	Bone, Egg Timer
25	Rat	59	Super Shed
26/27	McRat	60	Tug
28	Bread Loaf	66	Shed, Ying Ying
31/0	Toffee Apple (due to the power handle)	67	Skip
		68	Cat
31	Goyle, Brian, Ped	70	Ugly Duckling
33	Crompton, Shredder,	73	ED, Box, Shoebox
33/1	Bagpipe	76	Tommy
33/2	Slim Jim	81/85	Roarer
35	Hymek	86	Can
37	Tractor, Growler, Syphon	87	Van
37/9	Slug	88	Flymo, Sparky Cat
40	Whistler, Bucket	89	Badger, Aardvark
42	Warship	90	Skoda
43	HST, Flying Banana, Zing, Tram	91	Lada, Hairdryer
44-46	Peak, Wagon	92	Dyson
47	Duff, Brush, Spoon		

Right: Class 50s are popular amongst enthusiasts, with 18 of the 50 examples built now in preservation. They have at least four nicknames and of these, "Hoover", is the most widely used, the name being a nod to the sound made by the locos' original air intake fans. Class 50 bashers tend to refer to 50s as "Vacs", which is an evolution of the Hoover name. They have been called various other names too! 50003 is seen near Woking on the 12.20 Exeter–Waterloo on 6 May 1989.

APPENDIX 3 –
IMPRESSIVE BASHING ACHIEVEMENTS

Bashing is a never ending pursuit and there will always be locos that evade the basher, or were withdrawn before one got started. Some people have had a pretty good go at it though and racked up incredible numbers of haulages and mileages in the process. The table below lists some of these, which are understood to have been achieved by various committed individuals. There are some phenomenal mileages, at which the mind boggles when calculating the number of hours or days of travel these must have required. None of the achievements listed can be proved, they are simply claims that have been made by various individuals from a variety of sources, some on behalf of others who are no longer with us. There are a number of gaps needing to be filled in on this subject and any readers who can add to or correct the details below are encouraged to contact the author via the contact details at the beginning of this book, for updates and inclusion in any subsequent edition.

For each class, the number of locos built and the dates between which they operated is shown – these exclude any periods of departmental use, because whilst this extended the life of some locos, it was not for revenue earning or passenger duties. The dates give an indication of how difficult it is to clear some classes, as one would need to have started very early in some cases. Some locos rarely, if ever, stray from a particular route or region, meaning that wide ranging or intensive localised travel is needed, depending on one's priorities. Bear in mind also that once steam heating had been isolated for large numbers of classes (such as 31, 37 and 47 by the mid-1980s), these were then much more difficult to track down.

Some of this data is very recent and the figures will increase in time. For example, the last Class 68 was only introduced two years ago and some of the class have hauled very few passenger trains. Consequently, only three people are known to have "cleared" the class, with a number of others very close to doing so and mileage exploits increasing by the month.

Class	Total Built	Years Introduced	Years Withdrawn	Highest number of Haulages	Highest Total Class Mileage	Highest Individual Loco Mileage
08	996	1952-1962	1967-Present	About 200		
14	56	1964-165	1967-1969	Few while in service		
15	44	1957-1961	1968-1971	Very few		
16	10	1958	1968	None?	Very rare on passenger trains	
17	117	1962-1965	1968-1971			
20	228	157-1968	1976-Present	216		
21/29	58	1958-1960	1963-1971	1		
22	58	1959-1962	1967-1972	33		
23	11	1959	1968-1971	Unknown		
24	151	1958-1961	1967-1982			
25	327	1961-1967	1971-1987	Over 200		
26	47	1958-1959	1972-1993	All 47		
27	69	1961-1962	1966-1989	All 69		
28	20	1958-1959	1967-1977	19		
31/0	19	1957-1962	1976-1980	8		
31/1&4	244	1959-1962	1975-Present	All 244	360,000	35,203
33	98	1960-1962	1964-Present	All 98	750,000	36,279

Class	Total Built	Years Introduced	Years Withdrawn	Highest number of Haulages	Highest Total Class Mileage	Highest Individual Loco Mileage
35	101	1961-1964	1971-1975	About 50		
37	309	1960-1965	1966-Present	308 (all except D6983)	At least 500,000	60,000
40	200	1958-1962	1967-1988	All 200		
41	5	1958-1959	1967	Unknown		
42	38	1958-1961	1968-1972	All 38		
43	33	1960-1962	1969-1971	All 33		
43 HST	197	1972-1982	1998-Present	All 197	665,756	34,382
44	10	1959-1960	1976-1980	All 10		
45	127	1960-1963	1977-1989	All 127	414,529.75	24,847.75
46	56	1961-1963	1977-1991	All 56	105,000	
47	512	1962-1968	1965-Present	508, possibly higher	Over 1 million	60,843
50	50	1967-1968	1987-1994	All 50	Over 1 million	73,000
52	74	1961-1964	1973-1977	All 74		
55	22	1961-1962	1980-1982	All 22	Hundreds of thousands	35,000

Below: Of the many times I travelled on loco-hauled trains between Leeds and Carlisle, I never had the pleasure of doing so behind a pair of 31s. On 25 March 1989, 31407 & 31446 worked the 06.34 Leeds–Carlisle and 10.45 return to Leeds. The locos pass Garsdale during the morning's return to Leeds. *Richard Allen*

Above: 43032 first transported the author on one of the regular cross-country services to Manchester on 5 August 1987. With a new lease of life ahead of it after its 2018 refurbishment (and with 43130 on the rear), 43032 relaxes at Aberdeen after working the 09.28 from Edinburgh on 2 July 2019.

Class	Total Built	Years Introduced	Years Withdrawn	Highest number of Haulages	Highest Total Class Mileage	Highest Individual Loco Mileage
56	135	1976-1984	1991-Present	All 135		
57	33	1998-2004	None yet			
58	50	1983-1987	1999-2011	All 50		
59	15	1985-1995	None yet	All 15		
60	100	1989-1993	None yet	All 100		
66	485	1998-2016	2001-Present	200+		
67	30	1999-2000	None yet	All 30		
68	34	2012-2017	None yet	All 34	176,900	23,300
70 Diesel	37	2009-2017	None yet	2		
71	24	1958-1960	1967-1977	All 24		
73	49	1962-1967	1972-Present	48 (all except E6027)	Tens of thousands	
73/9	13	2013-2015	None yet			
74	10	1967-1968	1976-1977	All 10		
76	58	1941-1953	1970-1981			
77	7	1953-1954	1968	All 7		
81	25	1959-1964	1968-1991	All 25		
82	10	1960-1962	1969-1987	All 10		
83	15	1960-1962	1975-1989	All 15		
84	10	1960-1961	1977-1980	All 10		
85	40	1961-1964	1983-1992	All 40		
86	100	1965-1966	1986-Present	All 100		Over 122,000
87	36	1973-1975	1989-Present	All 36		Also over 122,000
88	10	2015-2016	None yet	8		
90	50	1987	None yet	All 50	Over 1 million	Over 800,000
91	31	1988-1991	None yet	All 31	2.55 million	564,000
92	46	1993-1996	None yet			

Left: Some examples of the classes which dominated the railways in the 1980s continue to work on the main line today. Of these, 20007 is the oldest, beginning its working life in 1957; it is seen at Wareham (with Class 33 No. D6515 on the rear), before working the 14.41 to Norden on 11 May 2018.

Above: The 16 year old version of your author, complete with Def Leppard t-shirt (they were popular in the late 1980s when the album Hysteria was out) at Manchester Piccadilly shortly before 31408 (front) and 31450 depart with the 08.37 Great Yarmouth–Liverpool Lime Street on 24 June 1989.